T0325289

Cambridge Tracts in Theoretical Computer Science 56
Rippling: Meta-Level Guidance for Mathematical Reasoning

Rippling is a radically new technique for the automation of mathematical reasoning. It is widely applicable whenever a goal is to be proved from one or more syntactically similar givens. The goal is manipulated to resemble the givens more closely, so that they can be used in its proof. The goal is annotated to indicate which subexpressions are to be moved and which are to be left undisturbed. It is the first of many new search-control techniques based on formula annotation; some additional annotated reasoning techniques are also described in the last chapter of the book.

Rippling was developed originally for inductive proofs, where the goal was the induction conclusion and the givens were the induction hypotheses. It has proved applicable to a much wider class of problems: from summing series via analysis to general equational reasoning.

The application to induction has especially important practical implications in the building of dependable IT systems. Induction is required to reason about repetition, whether this arises from loops in programs, recursive data-structures, or the behavior of electronic circuits over time. But inductive proof has resisted automation because of the especially difficult search control problems it introduces, e.g. choosing induction rules, identifying auxiliary lemmas, and generalizing conjectures. Rippling provides a number of exciting solutions to these problems. A failed rippling proof can be analyzed in terms of its expected structure to suggest a patch. These patches automate so called "eureka" steps, e.g. suggesting new lemmas, generalizations, or induction rules.

This systematic and comprehensive introduction to rippling, and to the wider subject of automated inductive theorem proof, will be welcomed by researchers and graduate students alike.

Cambridge Tracts in Theoretical Computer Science 56

Editorial Board

Titles in the series

Rippling

Meta-Level Guidance for Mathematical Reasoning

ALAN BUNDY
University of Edinburgh

DAVID BASIN
ETH Zürich

DIETER HUTTER
DFKI Saarbrücken

ANDREW IRELAND
Heriot-Watt University

CAMBRIDGE
UNIVERSITY PRESS

Shaftesbury Road, Cambridge CB2 8EA, United Kingdom

One Liberty Plaza, 20th Floor, New York, NY 10006, USA

477 Williamstown Road, Port Melbourne, VIC 3207, Australia

314–321, 3rd Floor, Plot 3, Splendor Forum, Jasola District Centre, New Delhi – 110025, India

103 Penang Road, #05–06/07, Visioncrest Commercial, Singapore 238467

Cambridge University Press is part of Cambridge University Press & Assessment,
a department of the University of Cambridge.

We share the University's mission to contribute to society through the pursuit of
education, learning and research at the highest international levels of excellence.

www.cambridge.org
Information on this title: www.cambridge.org/9780521834490

First published 2005

A catalogue record for this publication is available from the British Library

ISBN   978-0-521-83449-0   Hardback

To our wives, Josie, Lone, Barbara, and Maria,
for their patience during this 10 year project.

Contents

Preface

Automated theorem proving has been an active research area since the 1950s when researchers began to tackle the problem of automating human-like reasoning. Different techniques were developed early on to automate the use of deduction to show that a goal follows from givens. Deduction could be used to solve problems, play games, or to construct formal, mathematical proofs. In the 1960s and 1970s, interest in automated theorem proving grew, driven by theoretical advances like the development of resolution as well as the growing interest in program verification.

Verification, and more generally, the practical use of formal methods, has raised a number of challenges for the theorem-proving community. One of the major challenges is induction. Induction is required to reason about repetition. In programs, this arises when reasoning about loops and recursion. In hardware, this arises when reasoning about parameterized circuits built from subcomponents in a uniform way, or alternatively when reasoning about the time-dependent behavior of sequential systems.

Carrying out proofs by induction is difficult. Unlike standard proofs in first-order theories, inductive proofs often require the speculation of auxiliary lemmas. This includes both generalizing the conjecture to be proven and speculating and proving additional lemmas about recursively defined functions used in the proof. When induction is not structural induction over data types, then proof search is also complicated by the need to provide a well-founded order over which the induction is performed. As a consequence of these complications, inductive proofs are often carried out interactively rather than fully automatically.

In the late 1980s, a new theorem-proving paradigm was proposed, that of *proof planning*. In proof planning, rather than proving a conjecture by reasoning at the level of primitive inference steps in a deductive system, one could reason about and compose high-level strategies for constructing proofs.

The composite strategy could afterwards be directly mapped into sequences of primitive inferences. This technique was motivated by studying inductive proofs and was applied with considerable success to problems in this domain. Proof planning is based on the observation that most proofs follow a common pattern. In proofs by induction, if the inductive step is to be proven, then the induction conclusion (the goal to be proved) must be transformed in such a way that one can appeal to the induction hypothesis (the given). Moreover, and perhaps surprisingly, this transformation process, called *rippling*, can be formalized as a precise but general strategy.

Rippling is based on the idea that the induction hypothesis (or more generally hypotheses) is syntactically similar to the induction conclusion. In particular, an image of the hypothesis is embedded in the conclusion, along with additional differences, e.g., x might be replaced by $x + 1$ in a proof by induction on x over the natural numbers. Rippling is designed to use rewrite rules to move just the differences (here "$+1$") through the induction conclusion in a way that makes progress in minimizing the difference with the induction hypothesis. In Chapter 1 we introduce and further motivate rippling.

From this initially simple idea, rippling has been extended and generalized in a wide variety of ways, while retaining the strong control on search, which ensures termination and minimizes the need for backtracking. In Chapter 2 we describe some of these extensions to rippling including the application of rippling to proving noninductive theorems.

In contrast to most other proof strategies in automated deduction, rippling imposes a strong expectation on the shape of the proof under development. As previously explained, in each proof step the induction hypothesis must be embedded in the induction conclusion and the conclusion is manipulated so that the proof progresses in reducing the differences. Proof failures usually appear as missing or mismatching rewrite rules, whose absence hinders proof progress. Alternatively, the reason for failure might also be a suboptimal choice of an induction ordering, a missing case analysis, or an over-specific formulation of the conjecture. Comparing the expectations of how a proof should proceed with the failed proof attempt, so-called *critics* reason about the possible reasons for the failure and then suggest possible solutions. In many cases this results in a patch to the proof that allows the prover to make progress. In Chapter 3 we describe how these proof critics use failure in a productive way.

Since rippling is designed to control the proof search using the restrictions mentioned above, it strongly restricts search, and even long and complex proofs can be found quickly. In Chapter 5 we present case studies exemplifying the abilities of rippling. This includes its successes as well as its failures, e.g., cases where the restrictions are too strong and thereby prohibit finding proofs. We also present examples outside of inductive theorem-proving where rippling is used as a general procedure to automate deduction.

The above-mentioned chapters introduce techniques, extensions, and case studies on using rippling in an informal way, and provide a good overview of rippling and its advantages. In contrast, in Chapters 4 and 6 we formalize rippling as well as extending it to a more general and powerful proof methodology. The casual reader may choose to skip these chapters on the first reading.

In Chapter 4 we present the formal theory underlying rippling. In the same way in which sorts were integrated into logical calculi at the end of the 1970s, rippling is based on a specialized calculus that maintains the required contextual information. The restrictions on embeddings are automatically enforced by using a specialized matching algorithm while the knowledge about differences between the hypothesis and the conclusion is automatically propagated during deduction. The explicit representation of differences inside of formulas allows for the definition of well-founded orderings on formulas that are used to guarantee the termination of the rippling process.

Rippling is a successful example of the paradigm of using domain knowledge to restrict proof search. Domain-specific information about, for example, the difference between the induction conclusion and the induction hypothesis, is represented using term annotation and manipulated by rules of a calculus. In Chapter 6 we generalize the idea of rippling in two directions. First, we generalize the kinds of contextual information that can be represented by annotation, and we generalize the calculus used to manipulate annotation. The result is a generic calculus that supports the formalization of contextual information as annotations on individual symbol occurrences, and provides a flexible way to define how these annotations are manipulated during deduction. Second, we show how the various approaches to guiding proof search can be subsumed by this generalized view of rippling. This results in a whole family of new techniques to manage deduction using annotations.

In addition to this book there is a web site on the Internet at

$$\texttt{http://www.rippling.org}$$

that provides additional examples and tools implementing rippling. We encourage our readers to experiment with these tools.

Acknowledgments

We are grateful to Rafael Accorsi, Serge Autexier, Achim Brucker, Simon Colton, Lucas Dixon, Jurgen Doser, Bill Ellis, Andy Fugard, Lilia Georgieva, Benjamin Gorry, Felix Klaedtke, Boris Köpf, Torsten Lodderstedt, Ewen Maclean, Roy McCasland, Fiona McNeil, Raul Monroy, Axel Schairer, Jan Smaus, Graham Steel, Luca Viganó and Jürgen Zimmer, who read previous versions of this book and contributed to the book by stimulating discussions and comments. An especial thanks to Ewen Maclean for help with LaTeX.

We thank our editor David Tranah for the offer to publish this book at Cambridge University Press and for his patience during its preparation.

Finally, we thank Berendina Schermers van Straalen from the Rights and Permissions Department of Kluwer Academic Publishers who kindly granted us the right to make use of former journal publications at Kluwer.

1

An introduction to rippling

1.1 Overview

This book describes *rippling*, a new technique for automating mathematical reasoning. Rippling captures a common pattern of reasoning in mathematics: the manipulation of one formula to make it resemble another. Rippling was originally developed for proofs by mathematical induction; it was used to make the induction conclusion more closely resemble the induction hypotheses. It was later found to have wider applicability, for instance to problems in summing series and proving equations.

1.1.1 The problem of automating reasoning

The automation of mathematical reasoning has been a long-standing dream of many logicians, including Leibniz, Hilbert, and Turing. The advent of electronic computers provided the tools to make this dream a reality, and it was one of the first tasks to be tackled. For instance, the Logic Theory Machine and the Geometry Theorem-Proving Machine were both built in the 1950s and reported in *Computers and Thought* (Feigenbaum & Feldman, 1963), the earliest textbook on artificial intelligence. Newell, Shaw and Simon's Logic Theory Machine (Newell *et al.*, 1957), proved theorems in propositional logic, and Gelernter's Geometry Theorem-Proving Machine (Gelernter, 1963), proved theorems in Euclidean geometry.

This early work on automating mathematical reasoning showed how the rules of a mathematical theory could be encoded within a computer and how a computer program could apply them to construct proofs. But they also revealed a major problem: *combinatorial explosion*. Rules could be applied in too many ways. There were many legal applications, but only a few of these led to a proof of the given conjecture. Unfortunately, the unwanted rule applications cluttered up the computer's storage and wasted large amounts of processing power, preventing the computer from finding a proof of any but the most trivial theorems.

What was needed were techniques for guiding the search for a proof: for deciding which rule applications to explore and which to ignore. Both the Logic Theory Machine and the Geometry Theorem-Proving Machine introduced techniques for guiding proof search. The Geometry Machine, for instance, used diagrams to prevent certain rule applications on the grounds that they produced subgoals that were false in the diagram. From the earliest days of automated reasoning research, it was recognized that it would be necessary to use *heuristic* proof-search techniques, i.e. techniques that were not guaranteed to work, but that were good "rules of thumb", for example, rules that often worked in practice, although sometimes for poorly understood reasons.

1.1.2 Applications to formal methods

One of the major applications of automated reasoning is to formal methods of system development. Both the implemented system and a specification of its desired behavior are described as mathematical formulas. The system can then be verified by showing that its implementation logically implies its specification. Similarly, a system can be synthesized from its specification and an inefficient implementation can be transformed into an equivalent, but more efficient, one. Formal methods apply to both software and hardware. The use of formal methods is mandatory for certain classes of systems, e.g. those that are certified using standards like ITSEC or the Common Criteria.

The tasks of verification, synthesis, and transformation all require mathematical proof. These proofs are often long and complicated (although not mathematically deep), so machine assistance is desirable to avoid both error and tedium. The problems of search control are sufficiently hard that it is often necessary to provide some user guidance via an interactive proof assistant. However, the higher the degree of automation then the lower is the skill level required from the user and the quicker is the proof process. This book focuses on a class of techniques for increasing the degree of automation of machine proof.

Mathematical induction is required whenever it is necessary to reason about repetition. Repetition arises in recursive data-structures, recursive or iterative programs, parameterized hardware, etc., i.e. in nearly all non-trivial systems. Guiding inductive proof is thus of central importance in formal methods proofs. Inductive proof raises some especially difficult search-control problems, which are discussed in more detail in Chapter 3. We show there how rippling can assist with these control problems.

1.1.3 Proof planning and how it helps

Most of the heuristics developed for guiding automated reasoning are local, i.e., given a choice of deductive steps, they suggest those that are most promising. Human mathematicians often use more global search techniques. They first form an overall plan of the required proof and then use this plan to fill in the details. If the initial plan fails, they analyze the failure and use this analysis to construct a revised plan. Can we build automated reasoners that work in this human way? Some of us believe we can. We have developed the technique of proof planning (Bundy, 1991), which first constructs a proof plan and then uses it to guide the search for a proof.

To build an automated reasoner based on proof planning requires:

- The analysis of a family of proofs to identify the common patterns of reasoning they usually contain.
- The representation of these common patterns as programs called *tactics*.
- The specification of these tactics to determine in what circumstances they are appropriate to use (their *preconditions*), and what the result of using them will be (their *effects*).
- The construction of a *proof planner* that can build a customized *proof plan* for a conjecture from tactics by reasoning with the tactics' specifications.

A proof planner reasons with *methods*. A method consists of a tactic together with its specification, i.e. its preconditions and effects. Methods are often hierarchical in that a method may be built from sub-methods. Figure 1.1 describes a method for inductive proofs, using nested boxes to illustrate a hierarchical structure of sub-methods, which includes rippling.

1.1.4 Rippling: a common pattern of reasoning

Rippling is one of the most successful methods to have been developed within the proof-planning approach to automated reasoning. It formalizes a particular pattern of reasoning found in mathematics, where formulas are manipulated in a way that increases their similarities by incrementally reducing their differences. By only allowing formulas to be manipulated in a particular, difference-reducing way, rippling prevents many rule applications that are unlikely to lead to a proof. It does this with the help of annotations in formulas. These

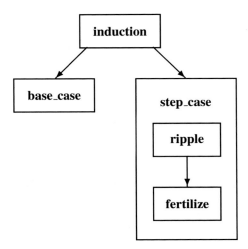

Figure 1.1 A proof method for inductive proofs. Each box represents a method. Arrows represent the sequential order of methods. Nesting represents the hierarchical structure of the methods. Note the role of rippling within the step case of inductive proofs. One base and one step case are displayed for illustration; in general, an inductive proof can contain several of each.

annotations specify which parts of the formula must be preserved and which parts may be changed and in what ways. They prevent the application of rules that would either change preserved parts or change unpreserved parts in the wrong way.

Rippling is applicable whenever one formula, the *goal*, is to be proved with the aid of another formula, the *given*. In the case of inductive proofs, the goal is an induction conclusion, and the given is an induction hypothesis. More generally, the goal is the current conjecture and the given might be an assumption, an axiom, or a previously proved theorem. Rippling attempts to manipulate the goal to make it more closely resemble the given. Eventually, the goal contains an instance of the given. At this point, the given can be used to help prove the goal: implemented by a proof method called *fertilization*.

To understand rippling, the following analogy may be helpful, which also explains rippling's name. Imagine that you are standing beside a loch[1] in which some adjacent mountains are reflected. The reflection is disturbed by something thrown into the loch. The mountains represent the given and their reflection represents the goal. The ripples on the loch move outwards in concentric rings until the faithfulness of the reflection is restored. Rippling is the movement of ripples on the loch: it moves the differences between goal and given to where they no longer prevent a match. This analogy is depicted in Figure 1.2.

[1] Rippling was invented in Edinburgh, so basing the analogy in Scotland has become traditional.

The mountains represent the given and the reflection represents the goal. The mountains are reflected in the loch.

The faithfulness of this reflection is disturbed by the ripples. As the ripples move outwards, the faithfulness of the reflection is restored.

In proofs, the rippling of goals creates a copy of the given within the goal. This pattern occurs frequently in proofs.

Figure 1.2 A helpful analogy for rippling.

1.2 A logical calculus of rewriting

In order to describe rippling we must have a logical calculus for representing proofs. At this point we need introduce only the simplest kind of calculus: the rewriting of mathematical expressions with rules.[1] This calculus consists of the following parts.

---

[1] We assume a general familiarity with first-order predicate calculus and build on that. An easy introduction to first-order predicate calculus can be found in Velleman (1994).

- The *goal* to be rewritten. The initial goal is usually the conjecture and subsequent goals are rewritings of the initial one.
- Some (conditional or unconditional) *rewrite rules*, which sanction the replacement of one subexpression in the goal by another.
- A procedure, called the *rewrite rule of inference*, that specifies how this replacement process is performed.

In this simple calculus, all quantifiers are universal. Section 4.1.2 gives a more formal account of rewriting.

Rewrite rules can be based on equations, $L = R$, implications $R \rightarrow L$, and other formulas. They will be written as $L \Rightarrow R$ to indicate the direction of rewriting, i.e. that L is to be replaced by R and not vice versa. Sometimes they will have conditions, $Cond$, and will be written as $Cond \rightarrow L = R$. We will use the single shafted arrow $\rightarrow$ for logical implication and the double shafted arrow $\Rightarrow$ for rewriting. We will usually use rewriting to reason backwards from the goal to the givens. When reasoning backwards, the direction of rewriting will be the inverse of logical implication, i.e. $R \rightarrow L$ becomes $L \Rightarrow R$.

To see how rewrite rules are formed, consider the following equation and implication.

$$(X + Y) + Z = X + (Y + Z) \tag{1.1}$$

$$(X_1 = Y_1 \wedge X_2 = Y_2) \rightarrow (X_1 + X_2 = Y_1 + Y_2). \tag{1.2}$$

Equation (1.1) is the associativity of $+$ and (1.2) is the replacement axiom for $+$. These can be turned into the following rewrite rules.

$$(X + Y) + Z \Rightarrow X + (Y + Z) \tag{1.3}$$

$$(X_1 + X_2 = Y_1 + Y_2) \Rightarrow (X_1 = Y_1 \wedge X_2 = Y_2). \tag{1.4}$$

The orientation of (1.3) is arbitrary. We could have oriented it in either direction. However, there is a danger of looping if both orientations are used. We will return to this question in Section 1.8. Assuming we intend to use it to reason from goal to given, the orientation of (1.4) is fixed and must be opposite to the orientation of implication.

In our calculus we will adopt the convention that bound variables and constants are written in lower-case letters and free variables are written in upper case. Only free variables can be instantiated. For instance, in $\forall x.\ x + Y = c$ we can instantiate Y to $f(Z)$, but we can instantiate neither x nor c.[1] The

---

[1] And nor can we instantiate Y to any term containing x, of course, since this would capture any free occurrences of x in the instantiation into the scope of $\forall x$, changing the meaning of the formula.

upper-case letters in the rewrite rules above indicate that these are free variables, which can be instantiated during rewriting.

We will usually present rewrite rules and goals with their quantifiers stripped off using the validity-preserving processes called *skolemization* and *dual skolemization*, respectively. In our simple calculus, with only universal quantification, skolemization is applied to rewrite rules to replace their universal variables with free variables, and dual skolemization is applied to goals to replace their universal variables with *skolem constants*, i.e. constants whose value is undefined.

The conditional version of the rewrite rule of inference is

$$\frac{Cond \rightarrow Lhs \Rightarrow Rhs \quad Cond \quad E[Rhs\phi]}{E[Sub]}.$$

Its parts are defined as follows.

- The usual, forwards reading of this notation for rules of inference is "if the formulas above the horizontal line are proven, then we can deduce the formula below the line". Such readings allow us to deduce a theorem from a set of axioms. However, we will often be reasoning backwards from the theorem to be proved towards the axioms. In this mode, our usual reading of this rewrite rule of inference will be: "if $E[Sub]$ is our current goal and both $Cond \rightarrow Lhs \Rightarrow Rhs$ and $Cond$ can be proven then $E[Rhs\phi]$ is our new goal".
- $E[Sub]$ is the goal being rewritten and Sub is the subexpression within it that is being replaced. Sub is called the *redex* (for *red*ucible *ex*pression) of the rewriting. $E[Sub]$ means Sub is a particular subterm of E and in $E[Rhs\phi]$ this particular subterm is replaced by $Rhs\phi$.
- The ϕ is a substitution of terms for variables. It is the most general substitution such that $Lhs\phi \equiv Sub$, where $\equiv$ denotes syntactic identity. Note that ϕ is only applied to the rewrite rule and not to the goal.
- $Cond$ is the condition of the rewrite rule. Often $Cond$ is vacuously true in which case $Cond \rightarrow$ and $Cond$ are omitted from the rule of inference.

For instance, if rewrite rule (1.3) is applied to the goal

$$((c + d) + a) + b = (c + d) + 42$$

to replace the redex $(c + d) + 42$, then the result is

$$((c + d) + a) + b = c + (d + 42).$$

1.3 Annotating formulas

Rippling works by annotating formulas, in particular, the goals and those occurring in rewrite rules. Those parts of the goal that correspond to the given are marked for preservation, and those parts that do not are marked for movement. Various notations have been explored for depicting the annotations. The one we will use throughout this book is as follows.

- Those parts of the goal that are to be preserved are written without any annotation. These are called the *skeleton*. Note that the skeleton must be a well-formed formula.
- Those parts of the goal that are to be moved are each placed in a grey box with an arrow at the top right, which indicates the required direction of movement. These parts are called the *wave-fronts*. Note that wave-fronts are *not* well-formed formulas. Rather they define a kind of *context*, that is, formulas with holes. The holes are called *wave-holes* and are filled by parts of the skeleton.

This marking is called *wave annotation*. A more formal account of wave annotation will be given in Section 4.4.2.

Wave annotations are examples of *meta-level* symbols, which we contrast with *object-level* symbols. Object-level symbols are the ones used to form expressions in the logical calculus. Examples are 0, $+$, $=$ and $\wedge$. Any symbols we use *outside* this logical calculus are meta-level. Annotation with meta-level symbols will help proof methods, such as rippling, to guide the search for a proof.

For instance, suppose our given and goal formulas are

Given: $a + b = 42$
Goal: $((c + d) + a) + b = (c + d) + 42,$

and that we want to prove the goal using the given. The a, $+b =$, and 42 parts of the goal correspond to the given, but the $(c + d) +$ part does not. This suggests the following annotation of the goal

$$(\boxed{(c + d) + a}^{\uparrow}) + b = \boxed{(c + d) + 42}^{\uparrow}.$$

This annotation process can be automated. Details of how this can be done will be given in Section 4.3.

Note the wave-holes in the two grey boxes. The well-formed formulas in wave-holes are regarded as part of the skeleton and not part of the wave-fronts. So the skeleton of the goal is $a + b = 42$, which is identical to the given. There are two wave-fronts. Both contain $(c + d) +$. Each of the wave-fronts has an upwards-directed arrow in its top right-hand corner. These arrows indicate the direction in which we want the wave-fronts to move: in this case outwards,[1] which is the default direction. In Chapter 2 we will see situations in which inwards movement is desirable.

1.4 A simple example of rippling

To illustrate rippling, consider the example in Section 1.3. Suppose the rewrite rules from Section 1.2 are available. Rule (1.3) can be used to rewrite the goal

$$((c + d) + a) + b = (c + d) + 42$$

in three different ways:

$$((c + d) + a) + b = c + (d + 42)$$
$$(c + (d + a)) + b = (c + d) + 42$$
$$(c + d) + (a + b) = (c + d) + 42 \tag{1.5}$$

but the first two of these are counterproductive. Only the rewriting to (1.5) moves us towards the successful use of the given: $a + b = 42$. The other two rewritings are examples of the kind of unwanted rule applications that would cause a combinatorial explosion in a more complex example.

Using rippling we can reject the two unwanted rewritings but keep the desired one. We first annotate each of them with respect to the given, $a + b = 42$:

$$(\boxed{(c + d) + a}^{\uparrow}) + b = \boxed{c + (d + 42)}^{\uparrow} \tag{1.6}$$

$$(\boxed{c + (d + a)}^{\uparrow}) + b = \boxed{(c + d) + 42}^{\uparrow} \tag{1.7}$$

$$\boxed{(c + d) + (a + b)}^{\uparrow} = \boxed{(c + d) + 42}^{\uparrow}. \tag{1.8}$$

Afterwards we compare each of them in turn with the original annotated goal

$$(\boxed{(c + d) + a}^{\uparrow}) + b = \boxed{(c + d) + 42}^{\uparrow}.$$

- In (1.6) the right-hand side wave-front changed in character, but is still in the same place with respect to the skeleton, i.e. it has not moved from where it was originally. From the viewpoint of rippling, things are no better.[2] This rewriting can be rejected as representing no progress.

---

[1] Or upwards, if we think of the formula as being represented by its parse tree, cf. Figure 1.3.
[2] In fact, as we will see in Section 2.1.3, things are actually worse.

- In (1.7) the left-hand side wave-front has changed in character, but is also still in the same place with respect to the skeleton. So this situation is similar to the previous one.
- In (1.8) the left-hand side wave-front has moved outwards, i.e. it is attached to the skeleton at a point outside where it was originally. From the viewpoint of rippling, things have improved. This rewriting can be welcomed as representing real progress.

In Section 4.7 we will make precise the concept of progress that we are appealing to informally above. We will give a well-founded measure that must be reduced by every rippling step. This measure will be based on the position of the wave-fronts within the skeleton. It will not only give us a basis for rejecting some rewrites as non-progressive or even regressive, it will also ensure the eventual termination of rippling. Most automated reasoning methods do *not* terminate; in general, the attempt to prove a conjecture may continue indefinitely with neither success nor failure. Termination of a method is a very desirable property, since it restricts the search space of the method to a finite size. It will also play a role in Chapter 3, where termination is used to detect failure, which starts a process that attempts to generate a patch.

We can now apply rewrite rule (1.4) to goal (1.8) and then annotate the result to check for progress

$$\boxed{c + d = c + d \wedge a + b = 42}^{\uparrow}.$$

We see that the single wave-front is now attached at the outermost point in the skeleton, i.e. it has moved outwards as far as it can. This represents real progress, in fact, as much progress as is possible with rippling, which now terminates with success.

If we write the three successive rewritings in sequence, we can see more clearly the rippling effect:

$$
\begin{array}{rcl}
(\,(c+d) + \boxed{a}^{\uparrow}\,) + b & = & \boxed{(c+d) + 42}^{\uparrow} \\
\boxed{(c+d) + (a+b)}^{\uparrow} & = & \boxed{(c+d) + 42}^{\uparrow} \\
\boxed{c+d = c+d \wedge a+b = 42}^{\uparrow}. &
\end{array}
$$

With each successive ripple, the wave-fronts get progressively bigger and contain more of the skeleton within their wave-holes. Eventually, the whole of the skeleton is contained within a single wave-front. Compare this with the picture of concentric ripples on a loch depicted in Figure 1.2. It may also help to see the same ripple with the skeletons represented as trees, depicted in Figure 1.3.

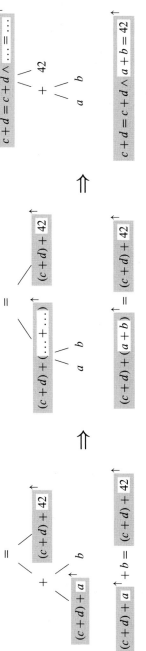

Figure 1.3 Rippling up the skeleton. The three trees show successive stages of rippling. Below each tree is the annotated formula it depicts. The trees consist of the tree of the skeleton, $a + b = 42$, with each wave-front attached to one of the nodes. A wave-front is attached to the node that is labeled by the top function symbol in its wave-hole, e.g. in the second tree the wave-hole of the wave-front $(c + d) + (a + b)$ is $a + b$, so this is attached to the $+$ node. Notice that the wave-fronts are attached to successively higher points of the tree, i.e. the wave-fronts are moving in the direction of the upwards arrow. Our wave-measure will be defined in terms of this upward movement of wave-fronts through the skeleton tree, i.e. the measure is lower the higher that the wave-fronts appear in the tree.

1.5 Using the given: fertilization

When rippling is complete, the goal contains an instance of the given. In our running example, the goal after rippling is

$$c + d = c + d \wedge \boxed{a + b = 42}^{\uparrow},\tag{1.9}$$

which consists of a wave-front, with the given, $a + b = 42$, as its wave-hole. To complete the proof it only remains to use the given to help prove goal (1.9). Since the given is assumed to be true, we can replace its instance in the goal with the boolean truth value $\top$. We can also drop the annotation, since it has served its purpose. In our example, this gives $c + d = c + d \wedge \top$, which is trivial to prove. This process is called *strong fertilization*. It can be regarded as a rewriting step where the rewrite rule is *given* $\Rightarrow \top$, e.g.

$$a + b = 42 \Rightarrow \top.$$

Sometimes it is not possible to ripple the wave-fronts completely outside the skeleton, but it is still possible to use the given to complete the proof. For instance, suppose in our running example that rewrite rule (1.4) was not available, so that the ripple was stuck at the goal

$$\boxed{(c + d) + (a + b)}^{\uparrow} = \boxed{(c + d) + 42}^{\uparrow}.$$

We can use the given, $a + b = 42$, as a rewrite rule left to right

$$a + b \Rightarrow 42$$

to rewrite the goal to

$$(c + d) + 42 = (c + d) + 42,$$

which again is trivial to prove. This is called *weak fertilization*.

1.6 Rewriting with wave-rules

In Section 1.4 we described rippling as a generate-and-test process: first rewrite the goal in all possible ways, then try to annotate the rewritings, then reject any rewritings that cannot be annotated or where the wave-fronts have not made progress. Generate-and-test procedures like this are usually inefficient. Greater efficiency can often be obtained by incorporating the testing stage into the generation stage, so that generation produces only results that pass the test.

This incorporation of testing within generation can be achieved in rippling by rewriting only with annotated rewrite rules, called *wave-rules*. The annotation of a wave-rule $Lhs \Rightarrow Rhs$ must have the following two properties.

Skeleton preservation: The skeletons of Lhs and Rhs must be the same.[1]

Measure decreasing: The annotation of the Rhs must represent progress over the annotation of the Lhs, i.e. the measure of the Rhs must be strictly less than that of the Lhs under a suitable well-founded ordering.

We will give formal definitions of these two properties in Chapter 4. In the meantime, an informal account is given in the caption to Figure 1.3. Furthermore, when a wave-rule is applied, its annotation must match the annotation in the redex. Under these three conditions we will see that successive goals are also skeleton preserving and measure decreasing. Since the measure is well-founded it cannot decrease indefinitely, so rippling always terminates (see Section 4.6.5). In practice, rippling terminates quite quickly, thus limiting the size of the search space and making automated reasoning more efficient than ordinary, unconstrained rewriting.

Rippling terminates when no further wave-rules apply. This is either because the goal is fully rippled or because rippling is stuck because no appropriate wave-rule is available. If the goal is fully rippled then the final goal must consist of a wave-front with an instance of the given in its wave-hole (cf. (1.9) above) or the wave-fronts have been entirely eliminated.

The rewrite rules from Section 1.2 provide examples of wave-rules. They can be annotated in several different ways, including the following:[2]

$$(\boxed{X + Y}^{\uparrow}) + Z \Rightarrow \boxed{X + (Y + Z)}^{\uparrow} \tag{1.10}$$

$$\boxed{X_1 + X_2}^{\uparrow} = \boxed{Y_1 + Y_2}^{\uparrow} \Rightarrow \boxed{X_1 = Y_1 \wedge X_2 = Y_2}^{\uparrow}. \tag{1.11}$$

Notice that they are both skeleton preserving and measure decreasing. The skeleton on each side of wave-rule (1.10) is $Y + Z$, and that on each side of wave-rule (1.11) is $X_2 = Y_2$. The wave-fronts are attached to the bottom of the skeleton on the left-hand sides and the top on the right-hand sides, showing that progress will be made when these rules are applied.

Consider the applications of wave-rule (1.10) to the annotated goal

$$(\boxed{(c + d) + a}^{\uparrow}) + b = \boxed{(c + d) + 42}^{\uparrow}.$$

---

[1] Later we will relax this requirement slightly.

[2] Wave-rule (1.11) is not annotated in the most general way. We will return to the issue of alternative annotations in Section 2.4.3.

- Wave-rule (1.10) *will not* apply to redex $\boxed{(c+d)+42}^{\uparrow}$ because its wave-fronts do not match those on the left-hand side of the wave-rule.
- Wave-rule (1.10) also *will not* apply to redex $\boxed{(c+d)+a}^{\uparrow}$, nor do its wave-fronts match those on the left-hand side of the wave-rule.
- Wave-rule (1.10) *will* apply to redex $\boxed{(c+d)+a}^{\uparrow})+b$ because its wave-fronts *do* match those on the left-hand side of the wave-rule.

So we see that by using wave-rules we do not even generate the unwanted rewritings of the goal.

1.7 The preconditions of rippling

Rippling is applicable whenever the current goal can be annotated so that its skeleton matches some given, i.e. some hypothesis, axiom, or previously established theorem. As we will see in Chapter 4, this is the case whenever the given is *embedded* in the goal. Intuitively, this means that the goal can be made identical to the given by "hiding" non-matching function symbols in the goal by placing them within wave-fronts. For example, $a+b$ can be embedded in $\boxed{s(a+b)}^{\uparrow}$ by hiding the subexpression $s(\ldots)$ in a wave-front. An algorithm calculating embeddings and annotating formulas will be given in Chapter 4.

Rippling works by repeated application of the sub-method *wave*. Each application of wave applies a wave-rule to the current goal to derive a new goal. Before it can be applied, the preconditions of the wave sub-method must be met. These preconditions are:

(i)   *The current goal has a redex that contains a wave-front.*
(ii) *There is a wave-rule whose left-hand side matches this redex.*
(iii) *If this wave-rule is conditional then its condition must be provable.*

In Section 2.2 we will investigate a new kind of rippling, which will require an additional precondition. In Chapter 3 we will see how different patterns of failure in these preconditions can suggest different kinds of proof plan patches.

For a wave-rule to rewrite a redex requires matching of the annotation as well as matching of the expressions. Suppose *Lhs* $\Rightarrow$ *Rhs* is to be applied to *redex*, then:

- *Any wave-fronts in Lhs must match identical wave-fronts in redex* (e.g. $\boxed{s(X)}^{\uparrow}$ *and* $\boxed{s(a+b)}^{\uparrow}$ *do match, but* $\boxed{s(X)}^{\uparrow}$ *and* $s(a+b)$ *do not*).

- *Any wave-fronts in redex must match either identical wave-fronts or free variables in Lhs*, e.g. $s(X)$ and $s(\boxed{s(n)}^{\uparrow})$ do match, but $\boxed{s(X)+Y}^{\uparrow}$ and $\boxed{s(a)+b}^{\uparrow}$ do not.

1.8 The bi-directionality of rippling

In Section 1.2 we used (1.1),

$$(X + Y) + Z = X + (Y + Z),$$

as a rewrite rule (1.3), left to right. But note that this equation (like all equations) is symmetric. Our decision to orient it left to right breaks this symmetry. Unfortunately, we are just as likely to need to use it right to left. Consider, for instance, the following problem, which is dual to the one in Section 1.3:

> Given: $a + b = 42$
> Goal: $a + (b + (c + d)) = 42 + (c + d)$.

To solve this problem we will need to use, not rewrite rule (1.3), but rule

$$X + (Y + Z) \Rightarrow (X + Y) + Z. \tag{1.12}$$

In an automated reasoning system it is dangerous to include both (1.3) and (1.12) as rewrite rules. One can invert the action of the other causing non-termination. For instance, they can generate the following endless sequence of rewritings

$$a + (b + (c + d)) = 42 + (c + d)$$
$$(a + b) + (c + d) = 42 + (c + d)$$
$$a + (b + (c + d)) = 42 + (c + d)$$
$$\vdots$$

amongst many others.

Rippling can avoid this problem. We can turn (1.1) into two wave-rules, one for each orientation. The wave annotation will prevent each of them undoing the action of the other and the termination of rippling will be preserved. These two wave-rules are

$$(\boxed{X+Y}^{\uparrow}) + Z \Rightarrow \boxed{X + (Y+Z)}^{\uparrow} \tag{1.13}$$

$$X + (\boxed{Y+Z}^{\uparrow}) \Rightarrow \boxed{(X+Y)+Z}^{\uparrow}. \tag{1.14}$$

Suppose the goal is annotated as

$$a + (\boxed{b} + (c+d)^{\uparrow}) = \boxed{42} + (c+d)^{\uparrow}.$$

Wave-rule (1.14) applies to this to produce

$$(\boxed{a+b}) + (c+d)^{\uparrow} = \boxed{42} + (c+d)^{\uparrow}.$$

Fortunately, wave-rule (1.13) does *not* apply to this goal; the left-hand side of the wave-rule does not match any redex in the goal because of mismatches between the wave-fronts. So the loop that occurred with conventional rewriting does not occur in rippling. More generally, the termination proof for rippling, which we will see in Section 4.6.5, still holds even when both (1.13) and (1.14) are present. The same holds for any other equation that can be annotated in both directions.

This ability of rippling to permit both orientations of a rewrite rule without the threat of non-termination, we call *bi-directionality*. It partially overcomes one of the major limitations of rewriting, namely that the orientation of rewrite rules causes incompleteness when the opposite orientation is required for a proof. In Section 5.1 we will see an example where bi-directionality is necessary.

This solution to incompleteness is only partial because, as we have seen in Section 1.6, rippling also prevents certain legal rewritings. The heuristic that underlies rippling ensures that the rewrites that are prevented are nearly always ones we did not want. However, sometimes rippling may prevent a rewrite that we do want. We will see examples and discussion of such situations in Chapter 5.

1.9 Proofs by mathematical induction

Many of the examples in this book will be drawn from inductive proofs. This is partly for historical reasons and partly because induction provides a rich source of examples. For completeness, we therefore include here a brief introduction to inductive proof.

Many people will have encountered some simple examples of induction in school mathematics. To prove a theorem for all natural numbers, 0, 1, 2, 3, . . . , by induction we consider a *base case* and a *step case*. In the base case we prove the theorem for 0. In the step case, we assume the theorem for n and prove it for $n + 1$. We can represent this inductive rule of inference as

$$\frac{\Phi(0) \quad \forall n{:}nat.\ (\Phi(n) \rightarrow \Phi(n + 1))}{\forall n{:}nat.\ \Phi(n)},$$

(1.15)

where *nat* is the type of natural numbers and n is the *induction variable*. In order to specify the types of variables, we have adopted a typed version of quantification, e.g. $\forall n{:}nat$ and $\exists n{:}nat$ both imply that n is of type *nat*. In the step case, given by the second premise, $\Phi(n)$ is called the *induction hypothesis* and $\Phi(n + 1)$ is called the *induction conclusion*.

1.9.1 Recursive data types

The natural numbers are an example of a *recursive data type*, i.e. a set of objects that are defined recursively. There are induction rules for every recursive data type. These include not just the natural numbers but also the integers, rationals, lists, trees, sets, etc.

To define a recursive data type we use a set of functions called *constructor functions*. The logician Giuseppe Peano (1858–1932) showed a simple way to do this for the natural numbers. He introduced two constructor functions: the constant 0 and the successor function s. The recursive construction rule for *nat* is as follows.

- 0 is a natural number.
- If n is a natural number then $s(n)$ is a natural number.

This gives a simple unary representation in which 1 is represented by $s(0)$, 2 by $s(s(0))$, 3 by $s(s(s(0)))$, etc. You may wonder why we could not stick to the conventional $0, 1, 2, \ldots, 42, \ldots$. The problem is that this introduces an infinite number of constants without any built-in connection between them. It can be made to work, but is messy. We could define a binary or decimal natural number recursive data type, but they would be a bit more complicated than the unary one, so we will stick with that.

Similarly, we can define a recursive data type for lists using the following constructors: the binary, infix function :: and the constant *nil*. The list $[a, b, c]$ can then be represented as $a :: (b :: (c :: nil))$. If τ is a type then let $list(\tau)$ denote the type of lists of elements of type τ.

Natural numbers and lists are examples of *free algebras*. That is, all syntactically distinct, variable-free terms are unequal, e.g. $s(s(0)) \neq s(0)$. Some recursive data types, however, are non-free. The integers, for instance, can be defined using the following constructors: the constant 0, the unary successor function, *succ*, and the unary predecessor function, *pred*.[1] Here 1 is represented by $succ(0)$ and -1 by $pred(0)$. However, 1 is also represented by $succ(succ(pred(0)))$. In general, $succ(pred(n)) = pred(succ(n))$. Induction can be used for proving theorems over both free and non-free algebras. A little more care must be taken when defining recursive functions over non-free algebras. It is impossible to avoid overlapping cases, i.e. giving two or more definitions for the same input. Therefore, it is necessary to prove that all the alternative definitions give the same output for the same input (cf. Sengler (1997) for mechanizing induction on non-free algebras).

1.9.2 Varieties of induction rule

There are induction rules for each recursive data type. Here, for instance, is an induction rule for lists:

$$\frac{\Phi(nil) \quad \forall h{:}\tau.\forall t{:}list(\tau).\ \Phi(t) \to \Phi(h :: t)}{\forall l{:}list(\tau).\ \Phi(l)}, \tag{1.16}$$

and here is one for integers:

$$\frac{\Phi(0) \quad \forall x{:}int.\ (\Phi(x) \to \Phi(succ(x))) \quad \forall x{:}int.\ (\Phi(x) \to \Phi(pred(x)))}{\forall x{:}int.\ \Phi(x)}.$$

Note that the integer rule requires two step cases. These, and rule (1.15), are examples of *structural induction*, where there is one inductive case for each constructor function.

Another dimension of variation is to base the induction on a different way of ordering the objects. For instance, we can have an induction on the natural numbers that goes up in steps of 2:

$$\frac{\Phi(0) \quad \Phi(s(0)) \quad \forall n{:}nat.\ \Phi(n) \to \Phi(s(s(n)))}{\forall n{:}nat.\ \Phi(n)}. \tag{1.17}$$

Note that this induction rule requires two base cases and that we have used the constructor function $s(n)$ instead of $n + 1$, as in rule (1.15).

Some theorems require quite exotic, custom-built induction rules. For instance, the standard proof that the arithmetic mean of a set of numbers is greater than or equal to the geometric mean uses the following rule:

$$\frac{\Phi(0) \quad \Phi(s(0)) \quad \forall n{:}nat.\ (\Phi(n) \to \Phi(2 \times n)) \quad \forall n{:}nat.\ (\Phi(s(n)) \to \Phi(n))}{\forall n{:}nat.\ \Phi(n)}.$$

---

[1] Note that, since integers are introduced here as a separate data type from the natural numbers, we have used different functions for the successor and predecessor constructor functions. Namely, *succ* and *pred* have been used for the integers, whereas *s* and *p* are used for the naturals.

Here the first step case goes up in multiples of 2 and the second one comes down in steps of 1.

Other dimensions of variation are to have multiple induction variables and multiple induction hypotheses. The following induction rule illustrates both of these:

$$\frac{\forall n{:}nat.\ \Phi(0, n) \quad \forall m{:}nat.\ \Phi(s(m), 0)}{\forall m{:}nat.\forall n{:}nat.\ (\Phi(m, n) \wedge \Phi(s(m), n) \wedge \Phi(m, s(n)) \rightarrow \Phi(s(m), s(n)))}{\forall m{:}nat.\forall n{:}nat.\ \Phi(m, n)}.$$

All the rules so far have used constructor functions in the induction conclusion. Our final variation is to use destructor functions in the induction hypothesis. Consider:

$$\frac{\forall n{:}nat.\ (n > 0 \rightarrow \Phi(p(n)) \rightarrow \Phi(n))}{\forall n{:}nat.\ \Phi(n)}, \tag{1.18}$$

where p is an example of a *destructor function*, i.e. it breaks down a recursive object rather than building one. The predecessor function for natural numbers is p, defined by

$$p(n) = \begin{cases} 0 & \text{if } n = 0 \\ m & \text{if } n = s(m) \end{cases}.$$

Rule (1.18) is the *destructor style* dual of the *constructor style* rule (1.15). Note that it has no base case. A base case will arise during most uses of rule (1.18), as a result of a case split on n.

All these, and more, induction rules can be derived from the general schema of Noetherian induction (also known as well-founded induction):

$$\frac{\forall x{:}\tau.\ (\forall y{:}\tau.\ y \prec x \rightarrow P(y)) \rightarrow P(x)}{\forall x{:}\tau.\ P(x)}, \tag{1.19}$$

where $\prec$ is some well-founded relation on the type τ, i.e. there are no infinite, descending chains of the form $\ldots \prec a_3 \prec a_2 \prec a_1$. It follows that $\prec$ is both non-reflexive and anti-symmetric.[1] Also, $\prec$ is a well-founded relation if and only if its transitive closure $\prec^+$ is well-founded. The Noetherian induction rules for $\prec$ and $\prec^+$ are inter-derivable. So, in practice, we often limit attention to transitive well-founded relations, which we call well-founded orders, since $\prec^+$ is a partial order.[2] In automated reasoning systems, Noetherian induction

---

[1] Otherwise $\ldots \prec a \prec a \prec a$ or $\ldots \prec b \prec a \prec b \prec a$.
[2] I.e., it is non-reflexive, anti-symmetric, and transitive.

is rarely used directly: rather, it is used to derive customized induction rules with specific base and step cases.

1.9.3 Rippling in inductive proofs

The step cases of inductive proofs provide an ideal situation for the use of rippling. The given is the induction hypothesis and the goal is the induction conclusion. In constructor-style induction rules, the wave-fronts arise from the functions that surround the induction variable. Rippling is not usually appropriate for the base cases of inductive proofs. Simple rewriting with the base equations of recursive definitions is often sufficient to prove these.

Consider, for instance, the use of induction to prove the distributive law of reverse over append:

$$\forall k{:}list(\tau),\ \forall l{:}list(\tau).\ rev(k <> l) = rev(l) <> rev(k), \qquad (1.20)$$

where *rev* is the list reversing function and $<>$ is the infix list appending function.

If we use induction rule (1.16) with induction variable k, then the base case is trivial:

$$rev(nil <> l) = rev(l) = rev(l) <> nil = rev(l) <> rev(nil).$$

The step case can be represented as the following rippling problem:[1]

Given: $rev(t <> l) = rev(l) <> rev(t)$

Goal: $rev(\boxed{h :: t}^{\uparrow} <> l) = rev(l) <> rev(\boxed{h :: t}^{\uparrow}).$

Wave-rules for this problem are provided by the recursive definitions of *rev* and $<>$, the associativity of $<>$, and the replacement axiom for $<>$:

$$rev(\boxed{H :: T}^{\uparrow}) \Rightarrow \boxed{rev(T) <> (H :: nil)}^{\uparrow}$$

$$(\boxed{H :: T}^{\uparrow}) <> L \Rightarrow \boxed{H :: (T <> L)}^{\uparrow}$$

$$X <> (\boxed{Y <> Z}^{\uparrow}) \Rightarrow \boxed{(X <> Y) <> Z}^{\uparrow}$$

$$(\boxed{X_1 <> X_2}^{\uparrow} = \boxed{Y_1 <> Y_2}^{\uparrow}) \Rightarrow \boxed{X_1 = Y_1 \wedge X_2 = Y_2}^{\uparrow}. \quad (1.21)$$

The full definitions of *rev* and $<>$, together with those of all other recursive functions used in this book, are given in Appendix 2.

---

[1] We will see in Section 2.2.2 that the *l*s in the goal can, in fact, take different values from those in the given, but this flexibility is not needed in this example.

With these wave-rules the rippling of the goal proceeds as follows:

$$rev(\boxed{h :: t}^{\uparrow} <> l) \;=\; rev(l) <> rev(\boxed{h :: t}^{\uparrow})$$

$$rev(\boxed{h :: t <> l}^{\uparrow}) \;=\; rev(l) <> (\boxed{rev(t)} <> (h :: nil)^{\uparrow})$$

$$\boxed{rev(t <> l)} <> (h :: nil)^{\uparrow} \;=\; \boxed{(rev(l) <> rev(t))} <> (h :: nil)^{\uparrow}$$

$$\boxed{rev(t <> l) = rev(l) <> rev(t)} \wedge (h :: nil) = (h :: nil)^{\uparrow}.$$

Note again how the wave-fronts ripple out with each successive ripple. The contents of their wave-holes grow until there is one wave-hole containing the whole of the given. Fertilization can now take place, leaving the remaining goal

$$\top \wedge (h :: nil) = (h :: nil),$$

which is trivial to prove.

Nearly all the step cases of inductive proofs follow some or all of this pattern of proof – even those based on non-structural and on destructor-style inductions. So rippling is just the right method for guiding the proofs of these step cases.

1.10 The history of rippling

The first person to use the term "rippling" was Aubin (1976). He pointed out that the application of recursive definitions during the step case of inductive proof usually caused the kind of rippling effect described by the loch analogy in Section 1.1.4. He called this effect "rippling out". Bundy (1988) turned this story on its head and suggested using annotations to *enforce* the rippling effect, rather than it be an emergent effect from another mechanism. The advantages of rippling over rewriting with recursive definitions were:

- Rippling directs the rewriting process towards an eventual fertilization by preventing rewrites that do not lead in this direction.
- Rippling applies to rewriting with lemmas, axioms, etc., as well as to recursive definitions.
- Rippling allows the bi-directional use of rewrite rules without sacrificing termination.

The term "fertilization" originated with Boyer and Moore (Boyer & Moore, 1979), who use the term "cross fertilization" to describe the substitution of equals for equals in an expression. Bundy (1988) identified its key role in the step case of an inductive proof as the target of the rippling process. Bundy, *et al.* (1990c) later distinguished weak and strong forms of fertilization.

Rippling was independently implemented both within Bundy's group at Edinburgh and by Hutter at Saarbrücken, the results being presented adjacently at the CADE-10 conference (Bundy *et al.*, 1990c; Hutter 1990). A longer version of Bundy *et al.* (1990c) appeared as Bundy *et al.* (1993). These papers also showed how the ideas could be generalized beyond the simple rippling-out pattern identified by Aubin. For instance, it was sometimes useful to ripple sideways and inwards (see Section 2.2). This led to the term "rippling-out" being replaced by the more general term "rippling". The development of rippling was part of the *proof planning* approach to automated theorem-proving (Bundy, 1988; 1991). Both rippling and fertilization were realized as proof methods.

Hutter's CADE-10 paper (Hutter, 1990), also started to develop a formal theory of rippling. Since then, various rival formal theories have been developed. Basin and Walsh (1996) gave formal definitions for wave annotations and the wave-measure. They used this to prove soundness and termination of rippling. They also developed algorithms for inserting wave annotation into formulas (Basin & Walsh, 1993). Hutter and Kohlhase developed an alternative account of wave annotation based on labeling terms with colors (Hutter & Kohlhase, 1997). This gave a very general account that could be applied to higher-order formulas and to embedded calls to rippling. Smaill and Green (1996) gave another general account of wave annotation, in which skeletons were defined as embeddings into formulas. This is also applicable to higher-order formulas.

Rippling is not guaranteed to succeed. It will fail if fertilization is not yet possible, but no wave-rule applies to the current goal. Unlike many other automated proof techniques, rippling carries a strong expectation of how the failed proof attempt should have proceeded. This expectation can often be used to analyze the cause of failure and suggest a patch to the proof. For instance, we can say a lot about the structure of any wave-rule that would allow the ripple to continue. This structure can be used to derive the missing wave-rule as a lemma. Alternatively, an existing wave-rule may almost apply and may be made applicable by either changing the form of induction, making a case split, or generalizing the goal. Automation of this proof analysis and patching process has been done by Ireland and Bundy (Ireland, 1992; Ireland & Bundy, 1996b).

Rippling has also been applied to help select an appropriate form of induction for a conjecture. The key idea is to choose a form of induction that will permit rippling to apply in the step case(s) of the inductive proof. A look-ahead is performed to see which wave-rules could apply in that step case and which choices of induction term and variable would maximize the chances of those wave-rules applying (Bundy *et al.*, 1989). This is called *ripple analysis*. Alternatively, a schematic induction rule can be applied in which the induction term is represented by a meta-variable.[1] This meta-variable is instantiated, by higher-order unification, during the application of wave-rules – effectively determining the induction rule as a side-effect of subsequent rippling (Gow, 2004; Kraan, *et al.*, 1996).

Rippling is well-suited to logical theories based on recursively defined functions: the movement of wave-fronts reflects the way in which function values are passed as the output of one function to the input of another as its argument. Unfortunately, in its original form, rippling did not work for theories based on recursively defined relations, as used, for instance, in logic programming languages such as Prolog (Kowalski, 1979). In such relational theories, values are passed from one relation to another via shared, existentially-quantified variables. A version of rippling, called *relational rippling*, has been developed for such relational theories (Bundy & Lombart, 1995).

Although rippling was originally developed for guiding the step cases of inductive proofs, it was discovered to be applicable to any situation in which a goal is syntactically similar to a given. Applications have been found to: summing series (Walsh *et al.*, 1992), limit theorems (Yoshida *et al.*, 1994), proofs in logical frameworks (Negrete, 1994), and equational theories (Hutter, 1997). The work on equational reasoning, in particular, offers the prospect of rippling playing a role in general-purpose theorem-proving. The rippling-based critics developed for inductive proof can also be applied to non-inductive problems. For instance, a critic originally developed to repair failed ripples by generalizing induction formulas, has been adapted to constructing loop invariants for verifying iterative, imperative programs (Ireland & Stark, 2001).

Rippling has been implemented at Edinburgh in the *C LAM* (Bundy *et al.*, 1990b), λ*C LAM* (Richardson *et al.*, 1998) and IsaPlanner (Dixon and Fleuriot, 2003) proof planners and at Saarbrücken in the INKA theorem prover (Hutter & Sengler, 1996). The Edinburgh *C LAM* implementation used wave-front context markers, to indicate where wave-fronts begin and end, but the later λ*C LAM* implementation uses a separate embedding record. The Saarbrücken INKA implementation uses symbol markers to indicate the status of each symbol: wave-front or skeleton. There has also been an implementation in NUPRL (Pientka & Kreitz, 1998). Early implementations predated the theoretical results described in Chapter 4 and adopted more ad hoc techniques, but the current implementations are theory-based.

---

[1] This meta-variable is so-called because it is not part of the logical calculus, in which object-level variables range over the elements of the domain, but is external to it and ranges over expressions of the calculus.

2

Varieties of rippling

The examples of rippling we met in Chapter 1 illustrate only part of the story. In this chapter we will discuss variations on the rippling theme. We will illustrate these variations with examples and in doing so we will gradually build up a more complete picture of what is possible using rippling.

2.1 Compound wave-fronts

Sometimes wave-fronts are compound, that is, they consist of more than a single function application. It may then be necessary to split the wave-front into parts and move it in several wave-rule applications.

2.1.1 An example of wave-front splitting

Consider, for instance, the proof of the theorem

$$\forall m{:}nat.\forall n{:}nat.\ even(m) \wedge even(n) \rightarrow even(m + n)$$

by the two-step induction rule (1.17). The step case produces the following rippling problem

> Given: $even(m) \wedge even(n) \rightarrow even(m + n)$
>
> Goal: $even(\boxed{s(s(m))}^{\uparrow}) \wedge even(n) \rightarrow even(\boxed{s(s(m))}^{\uparrow} + n),$

where the following wave-rules arise from the recursive definitions[1] of *even* and $+$

$$\boxed{s(X)}^{\uparrow} + Y \Rightarrow \boxed{s(X + Y)}^{\uparrow} \tag{2.1}$$

$$even(\boxed{s(s(X))}^{\uparrow}) \Rightarrow even(X). \tag{2.2}$$

---

[1] See Appendix 2 for the complete definitions of these functions.

Note that, in the *even* rule, there are no wave-fronts at all on the right-hand side. This is good, since the removal of wave-fronts always represents a decrease in the wave-measure and, therefore, progress towards fertilization.

The necessity for wave-front splitting arises in the application of wave-rule (2.1) to the redex $\boxed{s(s(m))}^{\uparrow} + n$. There is a mismatch between the wave-front on the left-hand side of the wave-rule and that in the redex. The solution to this is to split the compound wave-front in the redex into two nested wave-fronts, namely $s(\boxed{\boxed{s(m)}^{\uparrow}})^{\uparrow} + n$. Now the wave-front in wave-rule (2.1) matches the outer wave-front in the redex and the variable X matches $\boxed{s(m)}^{\uparrow}$. Applying the wave-rule gives $s(\boxed{\boxed{s(m)}^{\uparrow} + n})^{\uparrow}$. Notice that the outer wave-front has moved outwards and the inner one has stayed put. Wave-rule (2.1) can now be applied to the redex $\boxed{s(m)}^{\uparrow} + n$. This moves the inner wave-front outwards, giving $s(\boxed{\boxed{s(m+n)}^{\uparrow}})^{\uparrow}$. The complete sequence of rippling steps proceeds as follows:

$$even(\boxed{s(s(m))}^{\uparrow}) \wedge even(n) \rightarrow even(s(\boxed{\boxed{s(m)}^{\uparrow}})^{\uparrow} + n)$$

$$even(m) \wedge even(n) \rightarrow even(s(\boxed{\boxed{s(m)}^{\uparrow} + n})^{\uparrow})$$

$$even(m) \wedge even(n) \rightarrow even(s(\boxed{\boxed{s(m+n)}^{\uparrow}})^{\uparrow})$$

$$even(m) \wedge even(n) \rightarrow even(\boxed{s(s(m+n))}^{\uparrow})$$

$$even(m) \wedge even(n) \rightarrow even(m+n).$$

The goal is now reduced to the given and strong fertilization yields $\top$. Note that, in the penultimate step, we had to merge two nested wave-fronts into one compound one in order to apply wave-rule (2.2).

2.1.2 Maximally split normal form

To avoid having constantly to split and merge wave-fronts according to the wave-rules being applied, it is simplest to keep all wave-fronts in a maximally split normal form, i.e. all wave-fronts are one function symbol thick. However, it is unwieldy to write them in this maximally split form, so we will usually present them in a maximally merged form to reduce clutter. The reader needs to bear in mind that we are free to regard wave-fronts as split when this is needed for a wave-rule to apply.

Note that this splitting of wave-fronts is required only when more than one function symbol intercedes between the outer wave-front and the wave-hole. Compound terms that are wholly within the wave-front do not need to be split. To illustrate this distinction contrast the following two annotated terms:

$$rev(\ \boxed{(h_1 + h_2) :: \underline{t}}^{\uparrow}\ ) \qquad rev(\ \boxed{h_1 :: (\ \boxed{h_2 :: \underline{t}}^{\uparrow}\ )}^{\uparrow}\ ).$$

The left-hand wave-front has only one function symbol, namely ::, interceding between *rev* and the wave-hole t. Note that + does not intercede between them. In the right-hand wave-front, in contrast, two occurrences of :: intercede between *rev* and t. So here we do need to split the wave-front.

2.1.3 A promise redeemed

We are now in a position to redeem a promise made in footnote 2 of Section 1.4. We claimed there that the two unwanted rewritings:

$$(\ \boxed{(c+d) + \underline{a}}^{\uparrow}\ ) + b = \boxed{c + (\underline{d} + 42)}^{\uparrow}$$

$$(\ \boxed{c + (\underline{d} + a)}^{\uparrow}\ ) + b = \boxed{(\underline{c+d}) + 42}^{\uparrow}$$

were slightly worse than the original goal

$$(\ \boxed{(c+d) + \underline{a}}^{\uparrow}\ ) + b = \boxed{(\underline{c+d}) + 42}^{\uparrow}.$$

If we put all these wave-fronts into the maximally split normal form discussed in Section 2.1.2 then the original goal is unchanged but the two rewritings become:

$$(\ \boxed{(c+d) + \underline{a}}^{\uparrow}\ ) + b = \boxed{c + (\ \boxed{\underline{d} + 42}^{\uparrow}\ )}^{\uparrow}$$

$$(\ \boxed{c + (\ \boxed{\underline{d} + a}^{\uparrow}\ )}^{\uparrow}\ ) + b = \boxed{(\underline{c+d}) + 42}^{\uparrow}.$$

We can now see that each of the rewritings has replaced a single wave-front with a double one. We will see in Section 4.7 that increasing the number of wave-fronts at a point in the skeleton leads to an increase in the wave-measure. These rewritings would, therefore, be rejected as making matters worse.

2.1.4 Meta-rippling

Sometimes it is necessary to move wave-fronts without changing the underlying object-level formula. We call this *meta-rippling*, since the change is only visible in the wave annotation. To illustrate the need for this, consider the rippling problem

> Given: $even(n) \leftrightarrow odd(s(n))$
>
> Goal: $even(\boxed{s(s(\boxed{n}))}^{\uparrow}) \leftrightarrow odd(s(\boxed{s(s(\boxed{n}))}^{\uparrow})),$

where the following wave-rules are available:

$$even(\boxed{s(s(\boxed{X}))}^{\uparrow}) \Rightarrow even(X) \qquad (2.3)$$

$$odd(\boxed{s(s(\boxed{X}))}^{\uparrow}) \Rightarrow odd(X). \qquad (2.4)$$

Wave-rule (2.3) is immediately applicable to the left-hand side of the goal, but wave-rule (2.4) cannot yet be applied to the right-hand side.

To apply wave-rule (2.4) we must first ripple the goal to

$$even(\boxed{s(s(\boxed{n}))}^{\uparrow}) \leftrightarrow odd(\boxed{s(s(s(n)))}^{\uparrow})$$

using the wave-rule

$$s(\boxed{s(s(\boxed{X}))}^{\uparrow}) \Rightarrow \boxed{s(s(s(X)))}^{\uparrow}.$$

Note that this is skeleton preserving and measure decreasing, but causes no change at the object-level.

2.1.5 Unblocking rippling with simplification

It is sometimes necessary to interleave rippling with simplification steps. However, only wave-fronts should be simplified; simplifying skeletons will disrupt the embedding of the given into the goal (unless both are simplified in the same way). Simplification of wave-fronts is typically required to cleanup the application of one wave-rule and allow the next one to apply. We call this *unblocking*.

To illustrate the need for unblocking, consider the rippling problem

> Given: $x \times (y + z) = x \times y + x \times z$
>
> Goal: $\boxed{s(x)}^{\uparrow} \times (y + z) = \boxed{s(x)}^{\uparrow} \times y + \boxed{s(x)}^{\uparrow} \times z,$

where the following wave-rules are available:

$$\boxed{s(X)}^{\uparrow} \times Y \Rightarrow \boxed{X \times Y + Y}^{\uparrow} \tag{2.5}$$

$$A + (\boxed{B + C}^{\uparrow}) \Rightarrow \boxed{(A + B) + C}^{\uparrow} \tag{2.6}$$

$$(\boxed{A + B}^{\uparrow}) + C \Rightarrow \boxed{(A + C) + B}^{\uparrow} \tag{2.7}$$

$$\boxed{X_1 + Y}^{\uparrow} = \boxed{X_2 + Y}^{\uparrow} \Rightarrow X_1 = X_2. \tag{2.8}$$

Applying wave-rules (2.5) (twice), (2.6), and (2.7) to the goal produces

Given: $x \times (y + z) = x \times y + x \times z$

Goal: $\boxed{x \times (y + z) + (y + z)}^{\uparrow} = \boxed{((x \times y + x \times z) + y) + z}^{\uparrow}$,

but the goal is now blocked since wave-rule (2.8) cannot be applied. What is needed is to simplify the right-hand side wave-front with the associativity rewrite rule

$$(A + B) + C \Rightarrow A + (B + C).$$

After this simplification step, the rippling problem is transformed to

Given: $x \times (y + z) = x \times y + x \times z$

Goal: $\boxed{x \times (y + z) + (y + z)}^{\uparrow} = \boxed{(x \times y + x \times z) + (y + z)}^{\uparrow}$,

and wave-rule (2.8) can be applied. This reduces the goal to the given and allowing a trivial fertilization.

Unblocking has the potential to increase the rippling measure and, hence, undermine the termination of rippling. Fortunately, it is possible to define measures that both rippling and unblocking decrease, so that they can be interleaved and are still guaranteed to terminate. More details about termination can be found in Section 4.6.5.

2.2 Rippling sideways and inwards

Rippling identifies the differences between a goal and a given and moves these differences out of the way so that the goal contains an instance of the given. So far "out of the way" has meant moving wave-fronts outwards. In this section, we will consider other possible movements.

2.2.1 An example of sideways rippling

Suppose the given contains one or more free variables, e.g. K and L in

Given: $qrev(t, K <> L) = qrev(t, K) <> L$
Goal: $qrev(h :: t, k <> l) = qrev(h :: t, k) <> l,$

where *qrev* is a tail recursive list reversal function.[1] Note that the K and L in the given denote free variables, and the k and l in the goal denote constants. There is now an alternative "out of the way" place to move the differences between the given and the goal. If we can rewrite the goal to

$$qrev(t, (h :: k) <> l) = qrev(t, h :: k) <> l,$$

then fertilization can take place by instantiating K to $h :: k$ and L to l in the given. Instead of moving outwards, the differences have moved sideways.

This kind of situation arises frequently in inductive proofs. For example, as will be explained in Section 2.2.2, it arises when the induction formula has universally quantified variables, as skolemization turns these into free variables in the given (and skolem constants, via dual skolemization, in the goal). This is typically the case when proving facts about tail recursive functions such as *qrev* here. We can readily extend rippling to encompass this situation. We need to redefine the wave-measure to classify this kind of sideways movement as a decrease in the measure. We do this by extending the wave annotation to include both outwards and *inwards* wave-fronts. Inwards wave-fronts will be marked with a downwards arrow in the top right-hand corner, e.g. $qrev(T, \boxed{H :: L}^{\downarrow})$. The wave-measure will be extended so that the measure now decreases whenever either an inwards wave-front moves inwards or an outwards-directed wave-front turns into an inwards one.

However, not all inwards movement represents progress. The movement must be towards a term, such as k, which corresponds to a free variable, such as K, in the given. So that we can distinguish progress from regress, we will annotate such terms. We will call them *sinks* and annotate them as $\lfloor k \rfloor$.[2] An inwards-directed wave-front must have a sink in its wave-hole, i.e. it must have a target to ripple inwards towards.

Armed with this new annotation we return to our example. Suppose the following wave-rule is available from the tail recursive definition of *qrev*:

$$qrev(\boxed{H :: T}^{\uparrow}, L) \Rightarrow qrev(T, \boxed{H :: L}^{\downarrow}). \tag{2.9}$$

---

[1] See Appendix 2 for the complete definition. For an elementary explanation of tail recursion see Wikström (1987), Section 18.3.2.
[2] The marks are intended to suggest a kitchen sink with a drain at the bottom.

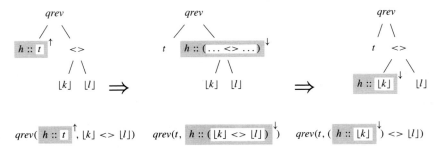

$$qrev(\boxed{h :: t}^{\uparrow}, \lfloor k \rfloor <> \lfloor l \rfloor) \quad qrev(t, \boxed{h :: (\lfloor k \rfloor <> \lfloor l \rfloor)}^{\downarrow}) \quad qrev(t, \boxed{h :: \lfloor k \rfloor}^{\downarrow}) <> \lfloor l \rfloor)$$

Figure 2.1 Rippling towards a sink. The three trees show successive stages of rippling. Wave-fronts are attached to the appropriate nodes of each tree. Below each tree is the annotated formula it depicts. The wave-front starts by moving sideways, but then moves downwards to immediately surround the sink.

This new kind of wave-rule is called a *transverse* wave-rule because it moves the wave-front sideways rather than up. Its wave-measure decreases from left to right because an outwards wave-front has turned into an inwards one.

With the aid of this wave-rule, the rippling in our example proceeds as follows:

$$qrev(\boxed{h :: t}^{\uparrow}, \lfloor k \rfloor <> \lfloor l \rfloor) = qrev(\boxed{h :: t}^{\uparrow}, \lfloor k \rfloor) <> \lfloor l \rfloor$$

$$qrev(t, \boxed{h :: (\lfloor k \rfloor <> \lfloor l \rfloor)}^{\downarrow}) = qrev(t, \boxed{h :: \lfloor k \rfloor}^{\downarrow}) <> \lfloor l \rfloor .$$

Using wave-rule (2.9) twice, the two outwards wave-fronts move sideways and become inwards wave-fronts. The right-hand wave-front has now immediately surrounded the sink $\lfloor k \rfloor$, but the left-hand wave-front is still some way above it. We need to ripple it inwards one more step. To do this we use an inwards version of wave-rule (1.21)

$$\boxed{H :: (T <> L)}^{\downarrow} \Rightarrow (\boxed{H :: T}^{\downarrow}) <> L$$

to produce

$$qrev(t, (\boxed{h :: \lfloor k \rfloor}^{\downarrow}) <> \lfloor l \rfloor) = qrev(t, \boxed{h :: \lfloor k \rfloor}^{\downarrow}) <> \lfloor l \rfloor .$$

The ripple of the left-hand wave-front is also represented as movement of wave-fronts on the tree of the skeleton in Figure 2.1. The wave-fronts have now reached their final destination. We can represent this by absorbing them into the sinks

$$qrev(t, \lfloor h :: k \rfloor <> \lfloor l \rfloor) = qrev(t, \lfloor h :: k \rfloor) <> \lfloor l \rfloor .$$

As required, strong fertilization is now possible. The free variable K is instantiated to the contents of the sink $h :: k$ and L to l. Note that we did not need to make use of the sink l as a target for wave-fronts.

2.2.2 Universal variables in inductive proofs

Opportunities to ripple into sinks are commonplace, for instance, in inductive proofs about tail recursive functions. The definitions of tail recursive functions will provide transverse wave-rules. Any universal variables in conjectures that are not used as induction variables will create sinks.

To understand this last claim, consider the abstract conjecture

$$\forall x{:}nat.\ \forall y{:}nat.\ \Phi(x, y).$$

Suppose induction rule (1.15) on x is applied to this. The step case will take the form

Induction hypothesis: $\forall y{:}nat.\ \Phi(x, y)$
Induction conclusion: $\forall y{:}nat.\ \Phi(s(x), y)$.

Notice that y is universally quantified both in the hypothesis and in the conclusion. The induction hypothesis is an assumption, so we can instantiate y in whatever way would assist the proof. The induction conclusion is a goal, so we must prove this goal for all values of y. The different status of y in each of these is realized by the processes of skolemization. The induction hypothesis is skolemized, which turns y into a free variable Y. The induction conclusion, being a goal, is dual-skolemized, which turns y into a skolem constant y_0, cf. Section 1.2. This constant y_0 is a sink. So the rippling problem is

Given: $\Phi(x, Y)$
Goal: $\Phi(\boxed{s(x)}^{\uparrow}, \lfloor y_0 \rfloor)$.

The rippling problem in Section 2.2.1 would arise, for instance, from the theorem

$$\forall x{:}list(\tau).\forall k{:}list(\tau).\forall l{:}list(\tau).\ qrev(x, k <> l) = qrev(x, k) <> l.$$

The production of sinks and free variables from non-inductive, universal variables is an aspect of inductive proofs that we have ignored, so far, in the interests of simplicity. For instance, the step case of (1.20) in Section 1.9.3 should have given rise to the following rippling problem

Given: $rev(t <> L) = rev(L) <> rev(t)$
Goal: $rev(\boxed{h :: t}^{\uparrow} <> \lfloor l \rfloor) = rev(\lfloor l \rfloor) <> rev(\boxed{h :: t}^{\uparrow})$,

instead of the rippling problem depicted in Section 1.9.3 without the free variables and sinks. We omitted this detail then in order not to complicate the story unnecessarily. In many proofs, as in this one, the sinks are not needed. We may then choose to omit the sink markers when including them would distract from the main story. The reader should be aware that rippling towards sinks may, nevertheless, be a possibility, i.e. a branch in the search space.

2.2.3 Revised preconditions for rippling

In Section 1.7 we described the preconditions of the ripple method and the wave sub-method. Now that we have extended rippling from rippling-out to rippling-sideways and rippling-in, we need to update these preconditions. In particular, we want to allow rippling-in only when this serves some purpose, namely to ripple a wave-front towards a sink. This suggests adding a fourth condition to the wave method: that we can only introduce an inwards wave-front if it has a sink in its wave-hole.

We start by repeating the first three preconditions from Section 1.7, then add the new condition.

  (i) The current goal has a redex that contains a wave-front.
 (ii) There is a wave-rule whose left-hand side matches this redex.
(iii) If this wave-rule is conditional then its condition must be provable.
(iv) *Any new inwards wave-front should have a sink in its wave-hole.*

2.2.4 Cancellation of inwards and outwards wave-fronts

In addition to sinks, there is another possible target for inwards-directed wave-fronts: they can sometimes be canceled with outwards-directed ones. Consider the following rippling problem arising in reasoning about permutations

> Given: $permute(t, t)$
> Goal: $permute(\boxed{h :: t}^{\uparrow}, \boxed{h :: t}^{\uparrow})$

in the presence of the following wave-rules:

$$X \in Z \rightarrow permute(\boxed{X :: Y}^{\uparrow}, Z) \Rightarrow permute(Y, \boxed{del(X, Z)}^{\downarrow}) \quad (2.10)$$

$$X = Y \rightarrow \boxed{del(X, \boxed{Y :: Z}^{\uparrow})}^{\downarrow} \Rightarrow Z. \quad (2.11)$$

Since $h \in h :: t$ and $h = h$, wave-rule (2.10) applies to the goal, followed by wave-rule (2.11), which yields the following rippling proof:

$$permute(\boxed{h :: t}^{\uparrow}, \boxed{h :: t}^{\uparrow})$$

$$permute(t, \boxed{del(h, \boxed{h :: t}^{\uparrow})}^{\downarrow}) \qquad (2.12)$$

$$permute(t, t).$$

As all wave-fronts are eliminated, fertilization with the given is possible, completing the proof.

Unfortunately, note that in line (2.12) there is an inwards wave-front whose wave-hole does *not* contain a sink. Precondition (iv) from Section 2.2.3 will, therefore, prevent this rippling step from being applied. However, what has happened in this ripple provides an alternative to the absorption of an inwards wave-front by a sink; the inwards wave-front has canceled with an outwards wave-front using wave-rule (2.11), in a form of destructive interference of wave-fronts. Such cancellation steps are quite common and our rippling preconditions must be extended to permit them. The new version of precondition (iv) is:

(iv)a *Any new inwards wave-front should have either a sink or an outwards wave-front in its wave-hole.*

Note that our precondition does not have to cover the case of a nested inwards wave-front since that will, in turn, contain an outwards wave-front or a sink.

2.3 Rippling-in after weak fertilization

Weak fertilization often leaves a residual goal to be proved. Rippling-in can often help in the proof of that goal.

2.3.1 An abstract example

To see why, consider the abstract rippling problem:

Given: $a(b(c(x))) = d(e(f(x)))$

Goal: $a(b(c(\boxed{s_1(x)}^{\uparrow}))) = d(e(f(\boxed{s_1(x)}^{\uparrow}))).$

Suppose the right-hand side of this goal can be fully rippled, but the left-hand side is blocked because of a missing wave-rule

$$a(b(\underbrace{\boxed{s_2(c(x))}^{\uparrow}}_{\textbf{blocked}})) = \boxed{s_3(d(e(f(x))))}^{\uparrow},$$

where s_2 and s_3 are the functions arising from rippling-out the s_1s. Weak fertilization can now take place on the right-hand side to give

$$a(b(s_2(c(x)))) = \boxed{s_3(a(b(c(x))))}^{\downarrow}.$$

The wave annotation on the left-hand side has been erased, since it has fully played its part. The wave annotation on the right-hand side has been retained, but the direction of the wave-front has been inverted in preparation for rippling this wave-front inwards. Suppose it is possible to ripple this wave-front past one occurrence of a, using the inwards-directed wave-rule

$$\boxed{s_3(a(Y))}^{\downarrow} \Rightarrow a(\boxed{s_4(Y)}^{\downarrow}) \qquad (2.13)$$

to produce

$$a(b(s_2(c(x)))) = a(\boxed{s_4(b(c(x)))}^{\downarrow}).$$

The two outermost occurrences of a can now be canceled to leave the simpler goal

$$b(s_2(c(x))) = s_4(b(c(x))). \qquad (2.14)$$

Note that precondition (iv) of the wave sub-method (see Section 2.2.4) prevents this use of rippling-in, so we need to redefine this precondition to be

(iv)b *If this wave application is prior to fertilization then any new inwards wave-front should have a sink or an outwards wave-front in its wave-hole.*

Now rippling-in is permitted after weak fertilization, even if the inwards wave-front does not contain a sink or an outwards wave-front.

2.3.2 Another way to look at it

Note that the goal (2.14) is an instance of the missing wave-rule, i.e. if the wave-rule

$$b(\boxed{s_2(Y)}^{\uparrow}) \Rightarrow \boxed{s_4(b(Y))}^{\uparrow} \qquad (2.15)$$

had been available in the first place then the ripple method would not have been blocked on the left-hand side. Furthermore, the outwards version of the inwards wave-rule (2.13)

$$a(\boxed{s_4(Y)}^{\uparrow}) \Rightarrow \boxed{s_3(a(Y))}^{\uparrow}$$

would have enabled that ripple to have been completed.

So if wave-rule (2.15) had been available, the following rippling steps would have been generated:

$$a(b(c(\boxed{s_1(x)}^{\uparrow}))) = d(e(f(\boxed{s_1(x)}^{\uparrow})))$$

$$a(b(\boxed{s_2(c(x))}^{\uparrow})) = \boxed{s_3(d(e(f(x))))}^{\uparrow} \qquad (2.16)$$

$$a(\boxed{s_4(b(c(x)))}^{\uparrow}) = \boxed{s_3(d(e(f(x))))}^{\uparrow} \qquad (2.17)$$

$$\boxed{s_3(a(b(c(x))))}^{\uparrow} = \boxed{s_3(d(e(f(x))))}^{\uparrow} \qquad (2.18)$$

$$a(b(c(x))) = d(e(f(x))),$$

at which point strong fertilization is possible. In the absence of wave-rule (2.15), weak fertilization has taken place at step (2.16), the wave-rule used at step (2.18) has been applied in reverse, and proving the missing wave-rule (2.15) has been left as a residual goal, i.e. essentially the same proof has taken place, but constructed in a different order.

2.3.3 A concrete example

This abstract pattern is realized in the proof of

$$\forall n{:}nat.\ half(n + n) = n,$$

where the recursive definitions of $+$ and *half* provide wave-rule (2.1) and

$$\boxed{s(half(X))}^{\downarrow} \Rightarrow half(\boxed{s(s(X))}^{\downarrow}). \qquad (2.19)$$

The step case generates the rippling problem:

Given: $half(n + n) = n$

Goal: $half(\boxed{s(n)}^{\uparrow} + \boxed{s(n)}^{\uparrow}) = \boxed{s(n)}^{\uparrow}.$

The goal can be rippled to

$$half(\boxed{s(n + \boxed{s(n)}^{\uparrow})}^{\uparrow}) = \boxed{s(n)}^{\uparrow},$$

but is then blocked. Fortunately, since the right-hand side is trivially fully rippled, weak fertilization can be applied yielding

$$half(s(n + s(n))) = \boxed{s(\,half(n + n)\,)}^{\downarrow}.$$

The right-hand side wave-front can now be rippled-in using wave-rule (2.19) and the outer functions canceled:

$$half(s(n + s(n))) = half(\,\boxed{s(s(n + n))}^{\downarrow})$$
$$s(n + s(n)) = s(s(n + n))$$
$$n + s(n) = s(n + n).$$

The result is an instance of the missing wave-rule

$$X + \boxed{s(Y)}^{\uparrow} \Rightarrow \boxed{s(X + Y)}^{\uparrow}.$$

In Chapter 3 we will see an alternative account in which the process above is carried out by a critic, called *lemma calculation*, for calculating the structure of missing lemmas.

2.4 Rippling towards several givens

It is sometimes possible to ripple towards several givens simultaneously. Both the need and the opportunity to do this arise, for instance, in inductive proofs with multiple induction hypotheses.

2.4.1 An example using trees

Consider the following conjecture about binary trees:

$$\forall t{:}tree(\tau).\ size(t) = count(nodes\_in(t)), \tag{2.20}$$

where $tree(\tau)$ is the type of binary trees whose nodes are labeled by elements of type τ; *size* returns the number of nodes in a tree, *nodes_in* returns a multi-set of all the nodes in a tree, and *count* computes the size of a multi-set. Suppose that the recursive data type $tree(\tau)$ is defined using the constructor functions *leaf* and *node*, and that the recursive data type $mset(\tau)$ is defined using the constructor functions *empty* and *insert*.

The step case of the structural induction on $tree(\tau)$ produces the rippling problem

Givens: $size(l) = count(nodes\_in(l))$
            $size(r) = count(nodes\_in(r))$

Goal: $size(\overline{node(n, \underline{l}, \underline{r})}^{\uparrow})$

$$= count(nodes\_in(\overline{node(n, \underline{l}, \underline{r})}^{\uparrow})).$$

Note that there are two givens: one corresponding to an induction hypothesis for the left subtree, and one for the right. Moreover, the wave-fronts now have two wave-holes: one corresponding to each of the givens. To handle this, we need to generalize the concept of skeleton: the skeleton is now defined as a set of formulas, e.g.

$$\{size(l) = count(nodes\_in(l)),\ size(r) = count(nodes\_in(l)),$$
$$size(l) = count(nodes\_in(r)),\ size(r) = count(nodes\_in(r))\}, \quad (2.21)$$

where each member of the set is constructed by choosing one of the wave-holes of each wave-front. Note that some of the members of the skeleton do not correspond to givens. We will see how to eliminate this problem by the use of colored wave-holes in Section 2.4.2.

We will proceed using the following six wave-rules. The first three come from the recursive definitions of *size*, *nodes_in* and *count*; the fourth comes from the distributive law of *count* over ∪; and the fifth and sixth come from the replacements axioms for *s* and +:

$$size(\overline{node(N, \underline{L}, \underline{R})}^{\uparrow}) \Rightarrow \overline{s(\underline{size(L)} + \underline{size(R)})}^{\uparrow}$$

$$nodes\_in(\overline{node(N, \underline{L}, \underline{R})}^{\uparrow}) \Rightarrow \overline{insert(N, \underline{nodes\_in(L)} \cup \underline{nodes\_in(R)})}^{\uparrow}$$

$$count(\overline{insert(E, \underline{S})}^{\uparrow}) \Rightarrow \overline{s(\underline{count(S)})}^{\uparrow}$$

$$count(\overline{\underline{X} \cup \underline{Y}}^{\uparrow}) \Rightarrow \overline{\underline{count(X)} + \underline{count(Y)}}^{\uparrow}$$

$$\overline{s(\underline{X})}^{\uparrow} = \overline{s(\underline{Y})}^{\uparrow} \Rightarrow X = Y$$

$$\overline{\underline{X_1} + \underline{X_2}}^{\uparrow} = \overline{\underline{Y_1} + \underline{Y_2}}^{\uparrow} \Rightarrow \overline{\underline{X_1 = Y_1} \wedge \underline{X_2 = Y_2}}^{\uparrow}. \quad (2.22)$$

Note that four of these wave-rules also have wave-fronts containing two wave-holes.

Armed with these wave-rules, rippling proceeds as follows:

$$size(\boxed{node(n,\, l,\, r)}^{\uparrow}) = count(nodes\_in(\boxed{node(n,\, l,\, r)}^{\uparrow}))$$

$$\boxed{s(\boxed{size(l) + size(r)})}^{\uparrow} = count(\boxed{insert(n,\, nodes\_in(l) \cup nodes\_in(r))}^{\uparrow})$$

$$\boxed{s(\boxed{size(l) + size(r)})}^{\uparrow} = \boxed{s(count(\boxed{nodes\_in(l) \cup nodes\_in(r)}^{\uparrow}))}^{\uparrow}$$

$$\boxed{\boxed{size(l) + size(r)}}^{\uparrow} = count(\boxed{nodes\_in(l) \cup nodes\_in(r)}^{\uparrow})$$

$$\boxed{size(l) + size(r)}^{\uparrow} = \boxed{count(nodes\_in(l)) + count(nodes\_in(r))}^{\uparrow}$$

$$\boxed{size(l) = count(nodes\_in(l)) \wedge size(r) = count(nodes\_in(r))}^{\uparrow}.$$

Rippling is now complete and strong fertilization is possible with *each* of the two givens.

2.4.2 Shaken but not stirred

The successive goals in the above ripple each have the same skeleton, namely the set of four equations given in (2.21): two members of this set correspond to the two givens, and the other two are superfluous. Rippling will often reduce the size of the skeleton set as it proceeds. This is fine as long as at least one member of the resulting set is identical to at least one of the givens. However, sometimes all members corresponding to the givens are eliminated and only superfluous members are retained. To see how this could happen, suppose the following wave-rule were available:

$$\boxed{X_1 + X_2}^{\uparrow} = \boxed{Y_1 + Y_2}^{\uparrow} \Rightarrow \boxed{X_1 = Y_2 \wedge X_2 = Y_1}^{\uparrow}, \quad (2.23)$$

which is a permuted variant of wave-rule (2.22). If this variant wave-rule were used instead of (2.22) then it would produce the goal

$$\boxed{size(l) = count(nodes\_in(r)) \wedge size(r) = count(nodes\_in(l))}^{\uparrow},$$

whose skeleton is

$$\{size(l) = count(nodes\_in(r)),\ size(r) = count(nodes\_in(l))\}.$$

Here, only the two superfluous members are retained, so fertilization is prevented. We need to stop this kind of thing happening; in the words of James Bond, we want skeletons to be "shaken, but not stirred".

One mechanism for doing this is to label skeletons in both goals and wave-rules. When applying wave-rules, we insist that these labels correspond, in addition to the other conditions on matching wave annotation. It is convenient to use color names for the labels. For instance, in our example ripple we may label the two givens "red" and "blue". Those parts of the goal's skeleton that correspond solely to the red (blue) given will also be labeled red (blue). Some parts of the goal's skeleton correspond to both givens; these will be labeled with both red and blue. The labels are formally represented as sets of color names, e.g. $\{r\}$, $\{b\}$ and $\{r, b\}$, where r stands for red and b for blue. With this additional color constraint, the skeleton of our running example is reduced to a doubleton, with both members corresponding to one of the givens, $\{size(l) = count(nodes\_in(l)), size(r) = count(nodes\_in(r))\}$. Figure 2.2 shows the *count* rippling problem using this notation. This figure is used again in Section 5.10 but with the skeleton portrayed in different colors, rather than the use of superscripts.

The skeletons of wave-rules are labeled with terms containing free variables ranging over sets. These labels are represented as superscripts below. The labeling is done in a way that preserves the integrity of the skeletons. The free variables are instantiated in the process of matching wave-rules to redexes. For instance, the wave-rules (2.22) and (2.23) are labeled as

$$\boxed{X_1{}^R + \boxed{X_2{}^B}^{\uparrow}} {}^{\uparrow} =^{R \cup B} \boxed{Y_1{}^R + \boxed{Y_2{}^B}^{\uparrow}} {}^{\uparrow} \Rightarrow \boxed{X_1 = Y_1{}^R \wedge \boxed{X_2 = Y_2{}^B}^{\uparrow}} {}^{\uparrow} \quad (2.22)$$

$$\boxed{X_1{}^R + \boxed{X_2{}^B}^{\uparrow}} {}^{\uparrow} =^{R \cup B} \boxed{Y_1{}^B + \boxed{Y_2{}^R}^{\uparrow}} {}^{\uparrow} \Rightarrow \boxed{X_1 = Y_2{}^R \wedge \boxed{X_2 = Y_1{}^B}^{\uparrow}} {}^{\uparrow}. \quad (2.23)$$

The goal is labeled as

$$\boxed{size(l)^{\{r\}} + \boxed{size(r)^{\{b\}}}^{\uparrow}} {}^{\uparrow}$$

$$=^{\{r,b\}} \boxed{count(nodes\_in(l))^{\{r\}} + \boxed{count(nodes\_in(r))^{\{b\}}}^{\uparrow}} {}^{\uparrow}.$$

Wave-rule (2.22) will apply to this with R instantiated to $\{r\}$ and B to $\{b\}$. However, wave-rule (2.23) will not apply since there is no consistent assignment of R and B to the labels of the goal. Thus the colored labels prevent an unwanted "stirring" of the skeletons.

2.4.3 Weakening wave-fronts

In Section 2.4.2 we used a multi-holed wave-rule based on the replacement axiom for $+$, (2.22). In Section 1.6 we used a single-holed version of this

Givens: $size(l)^{\{r\}}$ $=^{\{r\}}$ $count(nodes\_in(l))^{\{r\}}$

$size(r)^{\{b\}}$ $=^{\{b\}}$ $count(nodes\_in(r))^{\{b\}}$

Goal: $size(\boxed{node(n, \boxed{l^{\{r\}}}, \boxed{r^{\{b\}}})}^{\uparrow\,\{r,b\}})$

$=^{\{r,b\}}$ $count(nodes\_in(\boxed{node(n, \boxed{l^{\{r\}}}, \boxed{r^{\{b\}}})}^{\uparrow\,\{r,b\}}))$

Ripple:

$\boxed{s(\boxed{size(l)^{\{r\}}} + \boxed{size(r)^{\{b\}}})}^{\uparrow}$

$=^{\{r,b\}}\ count(\boxed{insert(n, \boxed{nodes\_in(l)^{\{r\}}} \cup \boxed{nodes\_in(r)^{\{b\}}})}^{\uparrow\,\{r,b\}})$

$\boxed{s(\boxed{size(l)^{\{r\}}} + \boxed{size(r)^{\{b\}}})}^{\uparrow}$

$=^{\{r,b\}}\ \boxed{s(\boxed{count(\boxed{nodes\_in(l)^{\{r\}}} \cup \boxed{nodes\_in(r)^{\{b\}}})}^{\uparrow\,\{r,b\}})}^{\uparrow}$

$\boxed{\boxed{size(l)^{\{r\}}} + \boxed{size(r)^{\{b\}}}}^{\uparrow}$

$=^{\{r,b\}}\ \boxed{count(\boxed{nodes\_in(l)^{\{r\}}} \cup \boxed{nodes\_in(r)^{\{b\}}})}^{\uparrow\,\{r,b\}}}^{\uparrow}$

$\boxed{\boxed{size(l)^{\{r\}}} + \boxed{size(r)^{\{b\}}}}^{\uparrow}$

$=^{\{r,b\}}\ \boxed{count(nodes\_in(l))^{\{r\}}} + \boxed{count(nodes\_in(r))^{\{b\}}}^{\uparrow}$

$\boxed{size(l) = count(nodes\_in(l))^{\{r\}} \wedge size(r) = count(nodes\_in(r))^{\{b\}}}^{\uparrow}$

Figure 2.2 Rippling with color labels. The goal is simultaneously rippled towards both the red and the blue given. The color coding keeps the two skeletons from becoming entangled. Note how large parts of the skeleton are shared initially but become separated into red and blue skeletons as the rippling proceeds. A version of this figure in color can be found in Figure 5.10.

wave-rule, (1.11). We reproduce them below:

$$\boxed{X_1 + X_2}^{\uparrow} = \boxed{Y_1 + Y_2}^{\uparrow} \Rightarrow \boxed{X_1 = Y_1 \wedge X_2 = Y_2}^{\uparrow} \qquad (2.22)$$

$$\boxed{X_1 + X_2}^{\uparrow} = \boxed{Y_1 + Y_2}^{\uparrow} \Rightarrow \boxed{X_1 = Y_1} \wedge \boxed{X_2 = Y_2}^{\uparrow}. \qquad (1.11)$$

We have seen occasions where each one of these wave-rules has been required, i.e. we need them both. Wave-rule (1.11) is called a *weakening* of

$$\boxed{X_1{}^{\{r\}} + X_2{}^{\{b\}}}^{\uparrow} =^{\{r,b\}} \boxed{Y_1{}^{\{r\}} + Y_2{}^{\{b\}}}^{\uparrow} \Rightarrow \boxed{X_1 = Y_1{}^{\{r\}} \wedge X_2 = Y_2{}^{\{b\}}}^{\uparrow} \quad (2.22)$$

$$\boxed{X_1 + X_2{}^{\{b\}}}^{\uparrow} =^{\{b\}} \boxed{Y_1 + Y_2{}^{\{b\}}}^{\uparrow} \Rightarrow \boxed{X_1 = Y_1 \wedge X_2 = Y_2{}^{\{b\}}}^{\uparrow} \quad (1.11)$$

Figure 2.3 Weakening wave-rules by removing colors. Wave-rule (1.11) can be formed from wave-rule (2.22) by removing the wave-holes around the red skeleton and merging this with the wave-front. Another weakening of (2.22) could be formed by removing the blue skeleton.

wave-rule (2.22). A weakening is formed by erasing the annotation around one or more wave-holes and merging their contents with the wave-front. This must be done so that the wave-rules are still skeleton preserving. In terms of color labels we are erasing the wave-holes containing one or more colors but preserving at least one other color. Figure 2.3 depicts this.

The implementation of weakened wave-rules involves a space/time trade-off. Either all the weakenings of a wave-rule can be stored and matched against any redex, or only full-strength wave-rules need be stored and weakening can be built into the application process. So, in the latter case, wave-rule (1.11) would not be stored but any required weakening would be applied to wave-rule (2.22) when this rule was applied.

2.5 Conditional wave-rules

We have seen how wave-rules are formed by annotating rewrite rules. Sometimes these rewrite rules have conditions and hence produce conditional wave-rules such as

$$X \neq H \rightarrow X \in \boxed{H :: T}^{\uparrow} \Rightarrow X \in T$$

$$H \in S \rightarrow (\boxed{H :: T}^{\uparrow}) \cap S \Rightarrow \boxed{H :: (T \cap S)}^{\uparrow}$$

$$X \neq H \rightarrow delete(X, \boxed{H :: T}^{\uparrow}) \Rightarrow \boxed{H :: delete(X, T)}^{\uparrow}, \quad (2.24)$$

where $\in$, $\cap$ and *delete* are the list membership, intersection, and element deletion functions, respectively.

Conditional wave-rules can be used for rippling provided their conditions are provable. Sometimes a wave-rules's conditions is unprovable. In order to apply it, it is then necessary to split the proof into cases. For instance, suppose the current goal is

$$delete(x, \boxed{h :: \boxed{(t <> l)}}^{\uparrow}) = delete(x, \boxed{h :: \boxed{t}}^{\uparrow}) <> delete(x, l).$$

For the ripple to continue we need to apply conditional wave-rule (2.24) to each side. The condition of both applications is $x \neq h$. Unfortunately, this condition is unprovable, so the ripple is blocked. To unblock it we can split the proof into two cases, $x = h$ and $x \neq h$. In the $x \neq h$ case, the wave-rules now apply and the ripple continues:

$$\boxed{h :: \boxed{delete(x, t <> l)}}^{\uparrow} = \boxed{h :: \boxed{delete(x, t)}}^{\uparrow} <> delete(x, l)$$

$$\boxed{h :: \boxed{delete(x, t <> l)}}^{\uparrow} = \boxed{h :: \boxed{delete(x, t) <> delete(x, l)}}^{\uparrow}$$

$$delete(x, t <> l) = delete(x, t) <> delete(x, l),$$

at which point strong fertilization completes this case.

The $x = h$ case can also proceed, but using the wave-rule for the complementary case

$$X = H \rightarrow delete(X, \boxed{H :: T}^{\uparrow}) \Rightarrow delete(X, T). \qquad (2.25)$$

Applying this wave-rule on each side produces

$$delete(x, t <> l) = delete(x, t) <> delete(x, l)$$

immediately, allowing strong fertilization to complete this case too.

The conditional wave-rules (2.24) and (2.25) form a complementary pair; each has the same left-hand side, but different right-hand sides and complementary conditions. If a case split is introduced on these complementary conditions, then each wave-rule enables continued rippling in one of the resulting cases. The existence of such a complementary pair of wave-rules is a strong heuristic suggesting a case split; it ensures that rippling will continue for at least one more step in both cases. More generally, to split into n cases we need n conditional wave-rules. If one of these conditional wave-rules is missing then the corresponding case will continue to be blocked. Fortunately, conditional wave-rules often arise from conditional recursive definitions (as here for $\in, \cap$ and *delete*) and the completeness of the definition provides rules for each of the cases.

Sometimes a complementary rewrite rule is available, but it is not a wave-rule. Consider, for instance, the complementary rules arising from the definition of $\in$:

$$X = H \rightarrow X \in H :: T \Rightarrow \top \tag{2.26}$$

$$X \neq H \rightarrow X \in \boxed{H :: T}^{\uparrow} \Rightarrow X \in T. \tag{2.27}$$

Rule (2.27) is a conditional wave-rule but rule (2.26) is not. Fortunately, we can still use these complementary rules to suggest a case split. However, due to the failure of skeleton preservation, one of these cases will cease to be a rippling problem. Consider, for instance, the goal

$$x \in \boxed{h :: (t <> l)}^{\uparrow} \leftrightarrow x \in \boxed{h :: t}^{\uparrow} \vee x \in l. \tag{2.28}$$

The complementary pair of rules (2.26) and (2.27) suggest a case split on $x = h$ and $x \neq h$. The second case continues the ripple. The first case, however, produces the goal

$$\top \leftrightarrow \top \vee x \in l,$$

which is no longer a rippling problem. Fortunately, it can be readily solved by conventional simplification methods, as is often so for such non-rippling cases.

Sometimes it is possible to replace a complementary set of conditional rules with a single wave-rule. For instance, the rules (2.26) and (2.27) can be replaced with the single wave-rule

$$X \in \boxed{H :: T}^{\uparrow} \Rightarrow \boxed{X = H \vee X \in T}^{\uparrow}. \tag{2.29}$$

Using such rules we can avoid a case split and, hence, any non-rippling cases. Applied to the goal (2.28), wave-rule (2.29) produces the ripple

$$x \in \boxed{h :: (t <> l)}^{\uparrow} \leftrightarrow x \in \boxed{h :: t}^{\uparrow} \vee x \in l$$

$$\boxed{x = h \vee x \in t <> l}^{\uparrow} \leftrightarrow (\boxed{x = h \vee x \in t}^{\uparrow}) \vee x \in l$$

$$\boxed{x = h \vee x \in t <> l}^{\uparrow} \leftrightarrow \boxed{x = h \vee (x \in t \vee x \in l)}^{\uparrow}$$

$$x \in t <> l \leftrightarrow x \in t \vee x \in l,$$

which is completed by strong fertilization.

In Chapter 3 we will see an alternative account in which the process above is carried out by a critic for suggesting case splits.

2.6 Rippling wave-fronts from given to goal

In all the rippling problems we have considered so far, wave annotation has appeared in the goal but not in the given. Rippling can also be applied to problems with wave annotation in the given. Such problems arise, for instance, from destructor-style inductions, e.g.

Given: $\boxed{p(l)}^{\uparrow} \times (m + n) = (\boxed{p(l)}^{\uparrow} \times m) + (\boxed{p(l)}^{\uparrow} \times n)$

Goal: $l \times (m + n) = (l \times m) + (l \times n),$

where p is the predecessor function for natural numbers introduced in Section 1.9.

It is tempting to treat such destructor-style rippling problems as entirely dual to the constructor-style ones we have considered up to now. That is, we might consider rippling the given and then fertilizing it with the goal. Unfortunately, this will not work. The problem is to prove the goal from the given; not the other way around. The given is an assumption, while the goal is the conjecture to be proved. It is not legal to use the conjecture to help prove the assumption.

Instead, somehow we must shift the wave annotation from given to goal and then ripple the goal, as before. There are a number of ways to do this. They all involve introducing wave annotation into the goal. Some of this annotation will correspond to the annotation already in the given. It will disappear if we re-embed the given into the goal. Some of the annotation will be new, and will form the basis for rippling the goal. By this process, the initial destructor-style problem will be transformed into a constructor-style problem.

The introduction of wave annotation into the goal can be effected by using annotated destructor-style definitions or similar rewrite rules. In our example we can use the conditional rule

$$X \neq 0 \rightarrow X \Rightarrow \boxed{s(p(X))}^{\uparrow} \qquad (2.30)$$

to split the rippling problem into two cases and rewrite the goal in the $l \neq 0$ case giving

Given: $\boxed{p(l)}^{\uparrow} \times (m + n) = (\boxed{p(l)}^{\uparrow} \times m) + (\boxed{p(l)}^{\uparrow} \times n)$

Goal: $\boxed{s(p(l))}^{\uparrow} \times (m + n) = (\boxed{s(p(l))}^{\uparrow} \times m) + (\boxed{s(p(l))}^{\uparrow} \times n).$

Now both the given and goal contain three occurrences of $p(l)$, so if we re-embed the given into the goal, these subexpressions will form part of the skeleton and not the wave-front, as previously. The rippling problem has thus been transformed to

Given: $p(l) \times (m + n) = (p(l) \times m) + (p(l) \times n)$

Goal: $\boxed{s(\,p(l)\,)}^{\uparrow} \times (m + n) = (\,\boxed{s(\,p(l)\,)}^{\uparrow} \times m) + (\,\boxed{s(\,p(l)\,)}^{\uparrow} \times n).$

Rippling of the goal can now proceed as if the original problem had been

Given: $l \times (m + n) = (l \times m) + (l \times n)$

Goal: $\boxed{s(l)}^{\uparrow} \times (m + n) = (\,\boxed{s(l)}^{\uparrow} \times m) + (\,\boxed{s(l)}^{\uparrow} \times n),$

as might have arisen from a constructor-style induction.

Note that rule (2.30) is *not* a wave-rule; it is skeleton preserving, but not measure decreasing. Great care must be taken with such measure-increasing rules since, applied indiscriminately, they can cause non-termination. The key here is to limit their application to the removal of wave annotation from the given, as in the above example.

2.7 Higher-order rippling

All our examples of rippling, so far, have been in the context of a first-order calculus. Can rippling also take place in a higher-order calculus, such as the λ-calculus? The answer is yes, but there are some additional complications that must be dealt with.

It is important to generalize rippling to higher-order calculi for two reasons. Firstly, many concepts from Mathematics and Formal Methods are naturally higher-order. Higher-order concepts include both induction and well-founded orderings, for instance. Secondly, as we will see in Chapter 3, to get the most out of rippling, we will want to use higher-order meta-variables to stand for unknown structure, for instance, in generalized conjectures, speculated lemmas, and induction terms.

So what are these additional complications in higher-order calculi? In contrast to first-order logic, the λ-calculus has an built-in equality relation on terms given by the α and β rules, and η-equality (see Figure 2.4).

Two terms are considered as equal if they are syntactically equal modulo these three laws. Let $[t]_{\alpha,\beta,\eta}$ be the class of all terms s that are equal to t modulo α, β, η-equality. Then, we need a concept of a skeleton that is independent of the selection of a representative in $[t]_{\alpha,\beta,\eta}$; i.e. $s \in [t]_{\alpha,\beta,\eta}$ should imply $skel(s) =_{\alpha,\beta,\eta} skel(t)$. It is easy to decide whether two terms are equal modulo the variable renaming of the α rule. However, the β rule and η equality create problems.

$$\frac{f[\lambda x\ e]}{f[\lambda y\ e\{y/x\}]} \qquad \frac{f[(\lambda x\ e)(t)]}{f[e\{t/x\}]} \qquad (\lambda y\ e)(y) = e$$

$$\alpha\ Rule \qquad\qquad \beta\ Rule \qquad\qquad \eta\ Equality$$

Figure 2.4 The laws of lambda calculus. Here f and e are expressions, $f[e]$ indicates a distinguished occurrence of e in f and e does not contain y.

The α rule allows us to switch bound variables; the β rule allows us to evaluate expressions; and the β rule and η equality are the two sides of a duality between function application and λ abstraction. The condition on e not containing y guards against accidental variable capture.

Consider the annotated term $(\lambda X.b)(\boxed{a})$. It is natural to consider a as its skeleton. However, if we apply the β rule to $(\lambda X.b)(\boxed{a})$, then it collapses to $\boxed{b}$ and the skeleton has vanished. Applying the β-rule has removed exactly the part of the term that contained the skeleton. The underlying problem is that the property of a symbol occurring in a term is not stable with respect to function composition. Although a occurs in all members of $[a]_{\alpha,\beta,\eta}$, it does not occur in all members of $[(\lambda X.b)a]_{\alpha,\beta,\eta}$. Below, we suggest two possible solutions to this problem.

Our first solution is to resign ourselves to the potential loss of the skeleton at each ripple step and, therefore, test for skeleton preservation after each step. To implement such a test, Smaill and Green (1996) developed the notion of *higher-order embeddings*. Informally, an expression A embeds into another B if A could be the skeleton when B was wave-annotated. We denote this as $A\ \mathcal{E}\ B$. Embeddings are discussed in more detail in Section 4.3.1, but this discussion is restricted to the first-order case. In the higher-order case, $\mathcal{E}$ is defined recursively as follows:

Base case: Each atomic expression B is embedded into itself: $B\ \mathcal{E}\ B$.

Application: A term A is embedded into an application $(B_1 B_2)$ if it is either embedded into one of its arguments (in this case the application would be some kind of wave-front) or A is itself an application $(A_1 A_2)$ and each A_i is embedded in B_i (in this case the application would be part of the skeleton).

Lambda abstraction: A is embedded into $(\lambda U.B)$ if either it is embedded into all instantiations $(\lambda U.B)C$ for all C (which interprets λ as a kind of wave-front), or A is itself an abstraction $(\lambda U.A')$ and $(\lambda U.A')C$ is embedded into $(\lambda U.B)C$ for all C (which interprets λ as part of a skeleton).

For instance, $a \; \mathcal{E} \; ((\lambda X.(fX))a)$ while it is not the case that $a \; \mathcal{E} \; ((\lambda X.b)a)$, which illustrates the necessity of checking the skeleton in each step. Thus, using this notion of higher-order embeddings we are able to implement a generate-and-test approach for rippling in a higher-order setting. In each step we have to check whether the skeleton is embedded in the derived formula.

However, compared to the first-order case, we lose some of the attractive properties of rippling: a first-order term $g(h(a, c), b)$, for instance, is written in the λ-calculus as $((g((ha)c))b)$. The relation between a function symbol and its arguments is blurred in the higher-order notation. Although b is not an argument of h, (hb) is embedded in $((g((ha)c))b)$. Thus, we lose expressiveness for specifying skeletons. Moreover, we are not able to trace individual symbol occurrences but have to search for corresponding symbol occurrences after each rippling step. As a consequence, we have to apply a wave-rule first in order to check whether it preserves the embedding of the skeleton in the manipulated term. In first-order logic, theorem 3 in Section 4.6.4 provides a sufficient condition to guarantee that the application of wave-rules preserves the skeleton. We do not need this theorem for the generate-and-test approach. It is trivially complete in the sense that it enables any rewriting provided that it preserves the embedding of the skeleton. Also, we are free to formulate arbitrary terms as skeletons regardless of the underlying types.

The above solution allows for the specification of arbitrary skeletons, which results in an expensive propagation of skeletons during rewriting (Hutter & Kohlhase, 2000). We now describe a second solution that restricts the possible skeletons in order to allow for the tracking of symbol occurrences. Consider the term $t \equiv ((\lambda X.A)B)$. The term B will only occur in all elements of $[((\lambda X.A)B)]_{\alpha,\beta,\eta}$ if X occurs in A (or B occurs itself in A). Thus, we cannot ignore $(\lambda X.A)$ completely when computing the skeleton of an annotated term like $(\;(\lambda X.A)\;\;B)$. To decide whether B is part of the skeleton, we need to know whether X occurs in A or not. Following the recursion scheme of higher-order embeddings, we want to compute a skeleton $skel((st))$ by composing the skeletons of the arguments $\{(s't')|s' \in skel(s), t' \in skel(t)\}$. Consequently, $skel(\;(\lambda X.A)\;)$ should return the set $\{(\lambda X.X)\}$ if X occurs in A and the empty set otherwise. We achieve this by considering λ and all occurrences of formal parameters like X as intrinsic parts of the skeleton, e.g. $skel(\;(\lambda X.f(X))\;) = \{(\lambda X.X)\}$. However, since $f(X)$ and X may have different types, we have to introduce additional type-conversion functions to guarantee well-typed skeletons and reduction rules to normalize the consecutive occurrences of these type-conversion functions. This second approach guarantees, in the ground case, that applying skeleton-preserving wave-rules will preserve the skeleton of the manipulated term.

However, there is one problem left: using higher-order variables we cannot guarantee that skeletons are stable with respect to instantiations. Consider for instance, a wave-rule like $\boxed{F(a)} \Rightarrow a$ with F being a higher-order variable. Instantiating F by a projection $\lambda x.b/F$ eliminates the skeleton a on the left-hand side and results in a non-wave-rule $\boxed{b} \Rightarrow a$. Hence, in general, the instance of a skeleton-preserving wave-rule is not itself skeleton preserving, but we have to check this property separately. Summing up, the second approach allows us to classify wave-rules in the same way as in the first-order case to restrict the number of suitable wave-rules when rewriting the given. Annotated higher-order matching between the given and the left-hand side of the wave-rule results in a set of annotated matchers. The wave-rule is instantiated by one of these matchers and the resulting instance is checked to ensure that it preserves the skeleton. If it does, we use this instance to rewrite the given, while otherwise we check the remaining set of matchers for admissible instantiations.

2.8 Rippling to prove equations

Rippling is applicable whenever a goal is to be proved with the aid of structurally similar givens. So far, the givens have been induction hypotheses, or similar assumptions, axioms, or lemmas. Rippling can be applied to prove equations by regarding the givens as the identities between the common skeletons of the two sides of the equation. Rippling then applies to both sides of the equation, moving away any differences from that common skeleton. This movement of the differences often results in an increase in the common skeleton, i.e. wave-fronts from each side are structurally similar; their change of position enables part of them to be absorbed into an increased skeleton. The process is repeated on any remaining differences, producing a steadily growing common skeleton until eventually all differences have been removed.

Consider, for example, the problem of proving

$$length(x <> y) = length(y <> x), \qquad (2.31)$$

using the distributive law of *length* over $<>$ and the commutativity of $+$. The differences and similarities between the two sides of (2.31) can be identified using the wave annotation

$$length(\boxed{x <> y}) = length(\boxed{y <> x}), \qquad (2.32)$$

where the common skeleton is $\{length(x),\ length(y)\}$. Next, rippling is used to move the wave-fronts out, so that the skeleton is contiguous. The distributivity of *length* over $<>$

$$length(\boxed{U <> V}) \Rightarrow \boxed{length(U)} + length(V)$$

applies to each side, to produce

$$\boxed{length(x)} + length(y) = \boxed{length(y)} + length(x) . \qquad (2.33)$$

Comparing both sides, we would like to increase the skeleton by the occurrences of $+$ on both sides, but the first argument of $+$ on the left-hand side corresponds to the second argument of $+$ on the right-hand side, and vice versa. Hence, in order to obtain a common skeleton on both sides, we have to use the commutativity of $+$ to permute the arguments of $+$ on one side of the equation yielding

$$\boxed{length(x) + length(y)} = \boxed{length(x) + length(y)} . \qquad (2.34)$$

Now the two sides of the equation are identical and the proof is complete. In wave annotation terms, $+$ has been moved from the wave-front to the skeleton, so that the new skeleton is

$$\{length(x) + length(y) = length(y) + length(x)\}. \qquad (2.35)$$

A more challenging example of equational rippling is given in Section 5.7.

Equational rippling in its most general form consists of two nested iterations. In the outer iteration, the common skeleton is incrementally enlarged by incorporating wave-fronts into the skeleton, until no wave-fronts remain and the two sides of the equation are identical. Each inner iteration prepares the way for one such a skeleton enlargement. Wave-fronts on each side are rippled into corresponding positions with respect to the common skeleton. Corresponding wave-fronts are then transformed until they are identical and can then be absorbed into the skeleton. Wave annotations are recalculated with respect to the new, enlarged, common skeleton, in preparation for the next step in the outer iteration.

This transformation of wave-fronts into the common skeleton is in contrast to standard rippling, in which the skeleton generally remains unchanged. The skeleton serves two purposes in this transformation process. Firstly, as in standard rippling, skeleton preservation restricts rewriting. Secondly, it allows us to reduce the problem of equalizing the two sides of the equation to the problem of equalizing corresponding wave-fronts. There is often a choice of the common skeleton. This choice is crucial, since it determines the way the problem is decomposed.

2.9 Relational rippling

Functions pass values from one to another by nesting one function inside the other and using the output of the inner function as the argument of the outer one. For instance, in $length(rev([1, 2, 3]))$ the output of $rev([1, 2, 3])$, $[3, 2, 1]$, is passed to $length$ as its argument. The version of rippling we have developed so far is designed for this environment of value passing by function nesting. So we might ripple $length(rev(\boxed{H :: T}^{\uparrow}))$ first to $length(\boxed{rev(T)} <> (H :: nil)^{\uparrow}))$ and then to $\boxed{s(length(rev(T)))}^{\uparrow}$, rippling the wave-front up through the nested functions.

However, in many application domains one works primarily with relations rather than functions. Examples include logic programming, reasoning about logic circuits represented as relations, and relational induction. In these domains, value passing is accomplished using shared, existentially quantified variables. For example, we might write $\exists R \exists N.\ rev([1, 2, 3], R) \wedge length(R, N)$, where, on execution, first R will become instantiated to $[3, 2, 1]$ and then N to 3. If we want rippling also to work in this case, then it needs to be modified. We call the new version of rippling *relational rippling*, and implement it using a variant of rippling called *rippling-past* (Bundy & Lombart, 1995).

In rippling-past, the wave-fronts move along a conjunction of literals, following the instantiation of the shared existentially quantified variables that carry the values being passed between conjuncts. Consider, for example, the relational versions of wave-rules in Figure 2.5 and the worked example of relational rippling in Figure 2.6.

In rippling-past we have to address a number of problems that do not arise in regular rippling:

(i) Conjunction is associative and commutative and existential quantification satisfies permutative laws (e.g., $\exists R \exists N.\phi$ is equivalent to $\exists N \exists R.\phi$) and permits some movement of quantifiers, as long as they preserve the scope of the quantification. We have to define skeleton preservation modulo these properties.

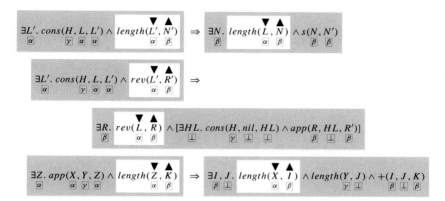

Figure 2.5 Relational versions of wave-rules. As usual, wave-fronts are represented by grey boxes and the wave-holes by the white areas within them. The arrows above variables act like Prolog modalities: down-arrows indicate an input and up-arrows an output. They impose a direction on the movement of wave-fronts and, thereby, enforce termination. The Greek letters in the boxes below arguments show the "real" value of these arguments and allow us to show skeleton preservation by first replacing the actual argument with the contexts of the box. So the skeleton of the first rule is $length(\alpha, \beta)$, etc.

(ii) As values are passed via shared variables, the names of variables in the skeletons are changed, violating skeleton preservation. We have to associate with each argument a constant name that is not changed by value passing, so that we can retain a concept of skeleton preservation. This annotation is done via boxes with Greek letters, as illustrated in Figures 2.5 and 2.6.

(iii) Function nesting gives a direction to the value passing: up through the nested functions. This direction is lost in value passing via shared existential variables; values can pass in either direction. We have, nevertheless, to impose some direction on relational wave-fronts, via the annotation, to ensure that the process terminates. This annotation is done via "modality" arrows, as illustrated in Figures 2.5 and 2.6.

(iv) An explicit initialization phase is needed to insert wave annotation into the step case of the proof. This transformation is illustrated in Figure 2.6.

(v) Rippling-past does not automatically move the wave-fronts onto the next part of the conjunction. It is necessary to have an explicit transformation to do this. This needs to regroup the conjunction, exploiting its properties of associativity and commutativity.

Relational rippling has been implemented in the *CIAM* system and successfully tested on a range of examples (Bundy & Lombart, 1995).

Conjecture: $\forall L, N. [\exists R. rev(L, R) \wedge length(R, N)] \leftrightarrow length(L, N)$

Step case:

$[\exists R. rev(l, R) \wedge length(R, n)] \leftrightarrow length(l, n)$,

$cons(h, l, \boxed{l'}) \vdash [\exists R'. rev(l', R') \wedge length(R', n')] \leftrightarrow length(l', n')$

Initialization:

Rippling-past:

Rippling-out:

Strong fertilization:

$\forall N. [\exists HL, N1. cons(h, nil, HL) \wedge length(HL, N1) \wedge +(N, N1, N')]$
$$\leftrightarrow succ(N, N')$$

Figure 2.6 An example of relational rippling. This shows the various stages of the proof of a conjecture using rippling-past and rippling-out. The wave annotation is as in Figure 2.6, except for the use of $\boxed{l'}$, which marks a candidate induction variable. Initialization sets up the initial wave annotation. Rippling-past then moves this annotation long the shared variable pathway and rippling-out and strong fertilization finish off the rippling in the conventional way.

2.10 Summary

In this chapter we have seen that the basic rippling story can be extended in many different ways. Wave-fronts can be compound objects that can be split, with the separate parts rippled differently. Rippling can send wave-fronts in different directions. It can move wave-fronts outwards, so that an instance of the given appears as a subterm; it can move the wave-fronts inwards towards a sink or in a phase of post-fertilization rippling; it can move wave-fronts across from given to goal. Complementary sets of rules can suggest splitting a proof into cases. It is possible to ripple simultaneously towards several givens, carefully preserving the skeletons corresponding to each one. It is also possible to extend rippling to higher-order logic and relational reasoning and to use it as the basis of a general procedure to automate equational reasoning.

We need to develop a uniform framework into which all these variants of rippling fit. This is the subject of Chapter 4. But first, we explore how to make productive use of the failure of rippling.

3

Productive use of failure

The process of proving theorems generally involves many false starts, whether performed using paper and pencil, or with machine assistance. But failed attempts often provide insight on how to make progress. Andrew Wiles expressed this phenomenon succinctly with regard to the famous patch of his proof of Fermat's Last Theorem: *"Out of the ashes seemed to rise the true answer to the problem"* (BBC, 1996). Such insights are sometimes referred to as *eureka steps* and the informality and reflective nature of the associated discovery process is common to many, whether they are students studying logic or professional mathematicians like Andrew Wiles. For instance, it is often the case that analysis of a proof failure suggests a generalization of the original conjecture that can be proved more easily. What is less common, however, is to find a symbolic method of proof that directly promotes this discovery process. We show in this chapter how rippling provides a basis for such a method.

3.1 Eureka steps

We begin by considering two of the key eureka steps that typically arise during the search for an inductive proof: conjecture generalization and lemma discovery. In terms of deduction,[1] both conjecture generalization and lemma discovery are underpinned by the *cut rule* of inference, which takes the form

$$\frac{\Gamma, \psi \vdash \phi \quad \Gamma \vdash \psi}{\Gamma \vdash \phi}.$$

---

[1] Our explanation assumes that a sequent calculus (Gentzen, 1934) is used as a deductive system. In other deductive systems (like Hilbert or natural deduction), the choice points described here are manifested in other ways, but the problem is the same: guessing and proving a new, valid formula, which may be syntactically unrelated to (e.g. is not a subformula of) the formula being proven.

We will make our usual assumption of a backwards style of proof construction. The cut rule allows us to exchange the goal of proving ϕ from Γ for two new sub-goals: (a) proving ϕ with the additional assistance of ψ, and (b) proving ψ from Γ; ψ is called the *cut formula*, and can be any formula. Thus, the cut rule introduces an infinite branching point in the search space. Additionally, there is a danger that ψ is not a consequence of Γ, so that the proof of sub-goal (b) will fail. This will occur, for instance, if a conjecture is over-generalized or an invalid lemma is applied.

In first-order logic, the cut rule is redundant, so these problems of infinite branching and over-generalization can be avoided. However, within first-order theories that include induction, the cut rule cannot be eliminated. For instance, cut elimination within theories of arithmetic is not possible. Moreover, as we will see below, the cut rule is frequently required for proving even quite simple theorems.

To illustrate these kinds of eureka steps, consider proving the following property of the *rotate* function

$$\forall t : list(\tau).\ rotate(length(t), t) = t. \tag{3.1}$$

The function *rotate* (defined in Appendix 2) takes a number n and a list l and returns the list built by removing the first n elements of l and concatenating them onto the end of l.

Conjecture (3.1) states that the rotation of a list by its length returns the identical list. Although intuitive and simple to state, this property is surprisingly hard to prove. A proof by mathematical induction requires first generalizing (3.1). The generalization step involves finding a conjecture that is provable by induction and logically implies (3.1). A suitable generalization here involves introducing an additional universally quantified variable, in particular

$$\forall t : list(\tau).\forall l : list(\tau).\ rotate(length(t), t <> l) = l <> t. \tag{3.2}$$

The advantage of this generalization is that an inductive proof of (3.2) gives rise to a stronger induction hypothesis than compared to an inductive proof attempt of (3.1). The discovery of (3.2) is an example of what we call *conjecture generalization*. In relationship to the cut rule of inference, (3.2) is the cut formula. Showing that (3.2) generalizes (3.1) requires establishing that (3.2) logically implies (3.1). This is relatively straightforward and is achieved by instantiating l to *nil* followed by the normalization of the antecedent, which gives rise to a tautology.

Once a generalized conjecture is discovered, however, it must also be proved. In the case of (3.2), a 1-step induction on the list t provides the basis for a proof. In the associated step case, however, the following two lemmas are required, each of which must also be proven by induction.

$$\forall x:list(\tau).\forall y:list(\tau).\forall z:list(\tau).\ (x <> y) <> z = x <> (y <> z) \qquad (3.3)$$

$$\forall y:\tau.\forall x:list(\tau).\forall z:list(\tau).\ x <> (y :: z) = (x <> y :: nil) <> z \qquad (3.4)$$

The discovery of (3.3) and (3.4) are examples of what we call *lemma discovery*. Relating this back to the cut rule of inference, both (3.3) and (3.4) represent cut formulas. Although these eureka steps appear to come from nowhere, we will return to this example in Section 3.7 and see how rippling provides the key to automating their discovery.

3.2 Precondition analysis and proof patching

The ability to discover eureka steps relies upon having a *strong expectation* of the shape the proof should take. As we have seen, rippling defines such an expectation for a particular family of proofs. In the remainder of this chapter, we show how rippling can be used to automate the discovery of certain kinds of eureka steps, including the ones illustrated above.

In proof planning terms, the "strong expectation" mentioned above corresponds to the explicit preconditions that define the applicability of the ripple method. Previously, we only considered situations where all the ripple preconditions succeeded, i.e.

(i)   The current goal has a redex that contains a wave-front.
(ii) There is a wave-rule whose left-hand side matches this redex.
(iii) If this wave-rule is conditional then its condition must be provable.
(iv) If this wave application is prior to fertilization then any new inwards wave-front should have a sink or an outwards wave-front in its wave-hole.

Here we systematically investigate how the *partial* success of these preconditions can be used to guide the search for proofs. In particular we will show that the tight constraints imposed by rippling mean that when rippling fails we have useful information about how to continue by *patching* the proof so that the failed constraints hold.

We present proof patching as a deductive process by which eureka steps are discovered. There are two kinds of proof patches that we use.

Modification patches: A patch may modify earlier steps within the proof attempt. This may be a local modification (i.e. involving a single goal) or a global modification involving subtrees of an existing proof tree.

Discovery patches: A patch may result in the discovery of a lemma that was not anticipated when the proof attempt began. In essence, we dynamically elaborate our knowledge of a theory during the course of planning a proof.

A key component of our proof-patching technique is the use of meta-variables, i.e. variables that range over the object-level syntax. We use meta-variables – typically higher-order meta-variables – to speculate missing structure during the planning of a proof. This technique has become known as *middle-out reasoning* (Bundy *et al.*, 1990a), as it can be used to delay decisions within a proof until the later proof steps have been worked out. Since meta-variables may appear within goals as well as rewrite rules, middle-out reasoning has similarities to *narrowing*, and in particular higher-order narrowing (Prehofer, 1994). We use meta-variables in two complementary ways: to speculate missing structure within goal formulas and lemmas.

In the case of a goal formula we speculate term structure that we believe will help the proof progress. For instance, suppose we are given a goal of the form

$$f(g(a, b), a).$$

Now suppose we have reason to believe that, in order to complete the proof, the second occurrence a should be generalized to a term that also contains c. This speculation can be expressed as the schematic goal

$$f(g(a, b), F_1(a, c)). \tag{3.5}$$

By "schematic" we mean that the goal contains a meta-variable, here the second-order variable F_1. In this example, $F_1(a, c)$ constitutes a term schema that can be thought of as representing a family of terms (namely, those substitution instances) that possibly contain both a and c. The expectation is that F_1 will become instantiated as a side-effect of planning a proof of (3.5). In terms of rippling, F_1 may be either part of the skeleton or a wave-front. If it is part of the skeleton then the patching process involves the discovery of a transformation to the original conjecture. On the other hand, if the missing term structure is a wave-front then the patching process involves the modification of earlier proof steps that will generate the required wave-front.

Now we turn to the use of meta-variables within the context of lemmas. Consider a goal of the form

$$f(g(a, b), \boxed{h(c)}^{\uparrow}). \tag{3.6}$$

This goal could be reduced if we had a wave-rule rule of the form

$$f(X, \boxed{h(Y)}^{\uparrow}) \Rightarrow \dots .$$

But how can the right-hand side be constructed? Within the context of rippling, the properties of structure preservation and termination can be used to guide the construction of a schematic right-hand side, e.g.

$$f(X, \boxed{h(Y)}^{\uparrow}) \Rightarrow \boxed{F_1(X, Y, f(\boxed{F_2(\underline{X})}^{\downarrow}, Y))}^{\uparrow} .$$

Here F_1 and F_2 are second-order meta-variables that denote *potential wave-fronts*, i.e. wave-fronts that may or may not exist. To be consistent with the representation of rippling, we refine our wave-front annotation so that a potential wave-front can be easily distinguished from an actual wave-front. This distinction is represented here by the use of a dotted box to denote a potential wave-front and an underline to denote its associated wave-hole. Again, the expectation is that the meta-variables will become instantiated as a side-effect of further proof planning.

The introduction of meta-variables eliminates our previous guarantee of termination with respect to proof search. To address this problem we interleave each wave-rule application with a fertilization step. Fertilization is now more than simple matching; it must deal with higher-order meta-variables. The strategy employed is defined such that fertilization instantiates a higher-order meta-variable to be a projection that preserves the skeleton term structure. Obviously, adopting such an *eager fertilization* strategy will lead to failures within the search space, e.g. it will give rise to candidate lemmas that are non-theorems. These unproductive paths are filtered-out using a counter-example finder.

Our proof-patching capability has been realized within the *proof critics* (Ireland, 1992) mechanism. Critics complement methods. While a method characterizes a family of proofs, a proof critic captures the patchable exceptions to that family. The ideas described in this chapter are based upon the development of proof critics for the ripple method (Ireland & Bundy, 1996a, 1996b, 1999).

3.3 Revising inductions

As described in Section 1.10, ripple analysis is a process for suggesting an appropriate form of induction to prove a conjecture. It works by looking-ahead to see which wave-rules could apply in the step case of an induction. However, ripple analysis only performs a 1-step look-ahead, and this is sometimes not sufficient. In this section we show how the analysis of a rippling failure can be used to revise an initial induction suggestion and thereby find a proof.

3.3.1 Failure analysis

We begin by exploring a failed proof attempt. Consider the conjecture

$$\forall t{:}list(\tau).\forall l{:}list(\tau).\ even(length(t <> l)) \leftrightarrow even(length(l <> t)), \quad (3.7)$$

where the predicate *even* holds for the even natural numbers and *length* computes the length of a list. The recursive definitions of *even* and *length*, given in Appendix 2, provide wave-rules including

$$even(\ \boxed{s(s(X))}^{\uparrow}\ ) \Rightarrow even(X) \qquad\qquad (3.8)$$

$$length(\ \boxed{X :: Y}^{\uparrow}\ ) \Rightarrow \boxed{s(length(Y))}^{\uparrow}. \qquad\qquad (3.9)$$

We also assume a lemma that relates *length* and $<>$, which provides a wave-rule of the form

$$length(X <> \boxed{Y :: Z}^{\uparrow}\ ) \Rightarrow \boxed{s(length(X <> Z))}^{\uparrow}.$$

In proving (3.7), rippling suggests a 1-step induction on the list t. Note that both t and l are equally good candidates, and our analysis works for either variable. We focus upon the step case, where the rippling problem is

Given: $even(length(t <> l)) \leftrightarrow even(length(l <> t))$

Goal: $even(length(\ \boxed{h :: t}^{\uparrow}\ <> l)) \leftrightarrow even(length(l <> \boxed{h :: t}^{\uparrow})).$

Initial rippling of the goal gives rise to

$$even(\ \underbrace{\boxed{s(length(t <> l))}^{\uparrow}}_{\text{blocked}}\ ) \leftrightarrow even(\ \underbrace{\boxed{s(length(l <> t))}^{\uparrow}}_{\text{blocked}}\ ).$$

No more wave-rules are applicable so the wave method fails to apply. This corresponds to the failure of the second precondition of rippling (see Sections 2.2.3 and 3.2):

(ii) There is a wave-rule whose left-hand side matches this redex.

Since the goal is not fully rippled, neither strong nor weak fertilization are applicable.

3.3.2 Patch: wave-front speculation

Our patch is based upon the observation that a partial match exists between the blocked wave-fronts and one of the available wave-rules. From this partial match, the structure of missing wave-fronts is calculated. The patch introduces the missing wave-fronts by revising our initial induction-rule selection. This analysis is implemented by dynamically constructing a schematic version of the blockage term in order to determine whether or not a partial match is possible. Here we need only consider one of the blockage terms in isolation; in general, this might not be the case. In this case, the schematic blockage term takes the form

$$even(\ s(\ F_1(length(Y\ <>\ F_2(Z)\ )) \ )\ ), \qquad (3.10)$$

where F_1 and F_2 are second-order meta-variables. Recall that dotted boxes denote *potential* wave-fronts, i.e. wave-fronts that may or may not exist. The left-hand side of wave-rule (3.8) unifies with (3.10), instantiating F_1 and F_2 to be $\lambda x.s(x)$ and $\lambda x.x$, respectively. This success suggests the need for an additional wave-front to overcome the blocked goal, i.e. an additional wave-front of the form $s(\dots)^{\uparrow}$ is required. The rippling-in of the composite wave-front (i.e. $s(s(\dots))^{\downarrow}$), is used to calculate the source of the missing wave-fronts.

$$even(\ s(s(\ length(t\ <>\ l))) \ ) \leftrightarrow even(\ s(s(\ length(l\ <>\ t))) \ )$$

$$even(\ s(\ length(\ h_2 :: t\ <>\ l\ )) \ ) \leftrightarrow even(\ s(\ length(l\ <>\ h_2 :: t\ )) \ )$$

$$even(\ s(\ length(\ h_2 :: t\ <>\ l)) \ ) \leftrightarrow even(length(l\ <>\ h_1 :: h_2 :: t\ ))$$

$$even(length(\ h_1 :: (\ h_2 :: t\ <>\ l)\ )) \leftrightarrow even(length(l\ <>\ h_1 :: h_2 :: t\ ))$$

$$even(length(\ h_1 :: h_2 :: t\ <>\ l)) \leftrightarrow even(length(l\ <>\ h_1 :: h_2 :: t\ )).$$

The result of the rippling-in is to suggest a 2-step instead of a 1-step induction. This revision gives rise to an induction hypothesis of the form

$$even(length(t\ <>\ l)) \leftrightarrow even(length(l\ <>\ t)), \qquad (3.11)$$

and a successful ripple as shown below.

$$even(length(\boxed{h_1 :: h_2 :: t}^{\uparrow} <> l)) \leftrightarrow even(length(l <> \boxed{h_1 :: h_2 :: t}^{\uparrow}))$$

$$even(length(\boxed{h_1 :: (\boxed{h_2 :: t}^{\uparrow} <> l)})) \leftrightarrow even(\boxed{s(length(l <> \boxed{h_2 :: t}^{\uparrow}))})$$

$$even(\boxed{s(length(\boxed{h_2 :: t}^{\uparrow} <> l))}) \leftrightarrow even(\boxed{s(s(length(l <> t)))}^{\uparrow})$$

$$even(\boxed{s(length(\boxed{h_2 :: t <> l}^{\uparrow}))}) \leftrightarrow even(\boxed{s(s(length(l <> t)))}^{\uparrow})$$

$$even(\boxed{s(s(length(t <> l)))}^{\uparrow}) \leftrightarrow even(length(l <> t))$$

$$even(length(t <> l)) \leftrightarrow even(length(l <> t)).$$

Strong fertilization with (3.11) completes the step-case proof. An alternative to the revision approach presented here is given in (Kraan *et al.*, 1996) where second-order function variables are used to delay the initial choice of induction rule. We will refer to terms constructed using function variables as *meta-terms*.

3.4 Lemma discovery

As explained in Section 3.1, lemmas play an important role in proving inductive theorems. Unless we know the lemmas required for a proof in advance the dynamic creation of missing lemmas will be required during the proof attempt. Here we focus upon how the partial success of rippling can be used to guide the discovery of missing lemmas.

3.4.1 Failure analysis

Consider the blocked goal

$$\dots \underbrace{\boxed{rev(t)} <> h :: nil}_{\textbf{blocked}}^{\uparrow} <> \lfloor l \rfloor \dots, \tag{3.12}$$

where *rev* is list reversal (defined in Appendix 2). The schematic version of the blockage term associated with (3.12) takes the form

$$\boxed{F_1(rev(Y))}^{\uparrow} <> X :: nil^{\uparrow} <> \lfloor l \rfloor.$$

Suppose that the search for a wave-rule that matches this schematic goal fails. This failure tells us that there are no wave-fronts that can be composed with $\boxed{\dots} <> X :: nil^{\uparrow}$ to enable further rippling. This rules out the induction revision proof patch described in Section 3.3. When partial matching fails, we use rippling to constrain the search for a missing wave-rule. Two complementary techniques for guiding the search for missing lemmas are presented below.

3.4.2 Patch: lemma speculation

To illustrate *lemma speculation*, consider the following blocked problem.

$$\text{Given:} \quad rev(rev(t) <> l) = rev(l) <> t \tag{3.13}$$

$$\text{Goal:} \quad rev(\underbrace{rev(t)\ \boxed{<> h :: nil}^{\uparrow} <> \lfloor l \rfloor}_{\text{blocked}})$$

$$= \underbrace{rev(\lfloor l \rfloor) <> \boxed{h :: t}^{\uparrow}}_{\text{blocked}} \tag{3.14}$$

Note that neither side of the goal equality is fully rippled. As outlined in Section 3.2, in such situations our approach to patching the complete failure of precondition (ii) involves constructing a schematic wave-rule. The expectation is that the planning of subsequent proof steps will generate instantiations for such meta-variables and consequently identify the missing lemma.

Our starting-point is the given blockage term. This forms the basis for the left-hand side of the missing wave-rule. In our example, this corresponds to

$$rev(l) <> \boxed{h :: t}^{\uparrow} \Rightarrow \dots .$$

The skeleton-preserving property (Section 1.3) of rippling provides us with a partial specification of the right-hand side of the missing wave-rule

$$\dots \Rightarrow \boxed{F_1(rev(l), h, t)}^{\downarrow} <> t.$$

Finally, we complete the construction of our alternative wave-rule schema by replacing common subterms in the skeleton by variables. This generalizes the applicability of the wave-rule and results in

$$X <> \boxed{Y :: Z}^{\uparrow} \Rightarrow \boxed{F_1(\underline{X}, Y, Z)}^{\downarrow} <> Z. \tag{3.15}$$

With the addition of this schematic wave-rule, the rippling of goal (3.14) is unblocked. That is, using wave-rule (3.15), goal (3.14) can be rewritten to

$$rev(\underbrace{\boxed{\boxed{rev(t)} <> h :: nil}^{\uparrow} <> \lfloor l \rfloor)}_{\textbf{blocked}} = \boxed{F_1(\underline{rev(\lfloor l \rfloor)}, h, t)}^{\downarrow} <> t.$$

As mentioned above, the expectation is that further rippling will incrementally instantiate F_1. Note that the direction of the potential wave-front coupled with the surrounding term structure constrains the search for applicable wave-rules.

The next step in the proof comes from the definition of *rev* that gives rise to a wave-rule of the form

$$rev(\boxed{rev(X)} <> \boxed{Y :: nil}^{\downarrow}) \Rightarrow rev(\boxed{Y :: X}^{\downarrow}). \tag{3.16}$$

The effect of this wave-rule on the right-hand side of the goal is

$$\ldots = \boxed{F_3(rev(\boxed{\boxed{F_2(rev(l), h, t) :: l}^{\downarrow}}), h, t)}^{\downarrow} <> t. \tag{3.17}$$

Here F_2 and F_3 result from the higher-order unification between $\boxed{F_1(rev(\lfloor l \rfloor), h, t)}^{\downarrow}$ and the left-hand side of (3.16). The result of this ripple is partially to instantiate wave-rule schema (3.15) to

$$X <> \boxed{Y :: Z}^{\uparrow} \Rightarrow \boxed{F_3(\boxed{X <> F_2(X, Y, Z) :: nil}^{\downarrow}, Y, Z)}^{\downarrow} <> Z.$$

Here, F_1 has been incrementally instantiated to

$$\lambda w.\lambda x.\lambda y.F_3(w <> F_2(w, x, y) :: nil, x, y).$$

In the case of goal (3.17), eager fertilization instantiates F_2 and F_3 with $\lambda w.\lambda x.\lambda y.x$ and $\lambda w.\lambda x.\lambda y.w$, respectively, giving rise to

$$rev(\underbrace{\boxed{\boxed{rev(t)} <> h :: nil}^{\uparrow} <> \lfloor l \rfloor)}_{\textbf{blocked}} = rev(\boxed{h :: l}^{\downarrow}) <> t. \tag{3.18}$$

The complete instantiation for (3.15) is now

$$X <> \boxed{Y :: Z}^{\uparrow} \Rightarrow \boxed{X <> Y :: nil}^{\downarrow} <> Z. \tag{3.19}$$

Once determined, the underlying lemma (3.4) must also be proved. In the case of (3.4), a simple structural induction is all that is required.

We now switch the focus to the left-hand side of (3.18). Note that (3.4) gives rise to a number of wave-rules including

$$\boxed{W <> X :: nil}^{\uparrow} <> Y \Rightarrow W <> \boxed{X :: Y}^{\downarrow}. \tag{3.20}$$

This is precisely the wave-rule that is required to unblock the left-hand side of (3.18):

$$rev(rev(t) <> \left\lfloor \boxed{h :: l}^{\downarrow} \right\rfloor) = rev(\left\lfloor \boxed{h :: l}^{\downarrow} \right\rfloor) <> t.$$

Note that we have used lemma (3.4) as a wave-rule in both orientations: left-to-right as wave-rule (3.19), and right-to-left as wave-rule (3.20). This further illustrates the bi-directionality of rippling (see Section 1.8).

Rippling is complete, and strong fertilization with (3.13) can now take place. In summary, the analysis of the initial failure to prove (3.14) is used to guide the discovery of lemma (3.4). To complete the proof of (3.14), the lemma is used in both orientations and rippling provides the necessary control to ensure termination.

3.4.3 Alternative lemmas

In the above discussion we side-stepped two choice points with respect to the construction of speculative wave-rules. First, there will generally exist a choice as to how much context should be included within the blockage term. This directly affects the structure of the resulting lemma. Second, there will generally exist a choice as to the positioning of the potential wave-fronts on the right-hand side of the wave-rule that is speculated. Using illustrative examples, we examine each choice point in turn.

To begin, consider the blocked goal

$$\ldots even(length(l <> (\boxed{h :: t}^{\uparrow}))) \ldots . \tag{3.21}$$

In the previous example we simply focused upon the blockage term that gave rise to the required lemma. Generally there will exist multiple blockage terms; in the case of (3.21) there are the following three alternatives.

$$even(length(\underbrace{l <> (\boxed{h :: t}^{\uparrow})}_{\textbf{blocked}})) \tag{3.22}$$

$$even(length(\underbrace{l <> (\boxed{h :: t}^{\uparrow})}_{\textbf{blocked}})) \tag{3.23}$$

$$even(length(\underbrace{l <> (\boxed{h :: t}^{\uparrow})}_{\textbf{blocked}})) \tag{3.24}$$

Based upon (3.22), a speculative wave-rule is constructed, namely

$$X <> \boxed{Y :: Z}^{\uparrow} \Rightarrow \boxed{F_1(\underline{X <> Z}, Y, Z)}^{\uparrow},$$

and rippling suggests

$$X <> \boxed{Y :: Z}^{\uparrow} \Rightarrow \boxed{Y :: (X <> Z)}^{\uparrow}$$

as an instantiation for this speculation. Our counter-example finder identifies this speculation as a non-theorem.

Now turning to (3.23), the associated speculative wave-rule takes the form

$$length(X <> \boxed{Y :: Z}^{\uparrow}) \Rightarrow \boxed{F_1(\underline{length(X <> Z)}, Y, Z)}^{\uparrow}.$$

Here, rippling suggests the instantiation

$$length(X <> \boxed{Y :: Z}^{\uparrow}) \Rightarrow \boxed{s(\underline{length(X <> Z)})}^{\uparrow}. \qquad (3.25)$$

This is precisely the wave-rule that is required to unblock (3.21), allowing us to ripple

$$\ldots even(length(l <> (\boxed{h :: t}^{\uparrow}))) \ldots$$

to

$$\ldots even(\boxed{s(\underline{length(l <> t)})}^{\uparrow}) \ldots .$$

Finally, in the case of (3.24), the speculation takes the form

$$even(length(X <> \boxed{Y :: Z}^{\uparrow})) \Rightarrow \boxed{F_1(\underline{even(length(X <> Z))}, Y, Z)}^{\uparrow}.$$

However, in this case the instantiation suggested by rippling again leads to a non-theorem, namely

$$even(length(X <> \boxed{Y :: Z}^{\uparrow})) \Rightarrow even(length(X <> Z)).$$

The observant reader may have noticed that an alternative induction actually is required, in addition to wave-rule (3.25), in order to fully ripple (3.21). As described in Section 3.3, our induction revision critic discovers the alternative induction.

We now consider the question of positioning potential wave-fronts on the right-hand side of speculative wave-rules. To illustrate the issue more clearly, we use the blockage term

$$a + \underbrace{\boxed{s(b)}^{\uparrow}}_{\text{blocked}}.$$

Based upon this blockage term, we can construct two wave-rules: one for rippling-out, and one for rippling-sideways, i.e.

$$X + \boxed{s(Y)}^{\uparrow} \Rightarrow \boxed{F_1(\underline{X+Y}, X, Y)}^{\uparrow} \tag{3.26}$$

$$X + \boxed{s(Y)}^{\uparrow} \Rightarrow \boxed{G_1(\underline{X}, Y)}^{\downarrow} + Y. \tag{3.27}$$

Note that in the previous examples we have ignored this choice point, focusing instead upon the form that produced the correct result. In this example, however, both speculations lead to success, i.e. in the case of (3.26) we get

$$X + \boxed{s(Y)}^{\uparrow} \Rightarrow \boxed{s(X+Y)}^{\uparrow}.$$

While in the case of (3.27) we get

$$X + \boxed{s(Y)}^{\uparrow} \Rightarrow \boxed{s(X)}^{\downarrow} + Y.$$

Combining longitudinal and transverse speculation gives rise to the following more general wave-rule (lemma) speculation

$$X + \boxed{s(Y)}^{\uparrow} \Rightarrow \boxed{F_1(\boxed{G_1(\underline{X}, Y)}^{\downarrow} + Y, X, Y)}^{\uparrow}.$$

To illustrate more concretely, a blocked term of the form $a + \boxed{s(s(b))}^{\uparrow}$ would give rise to wave-rules that include

$$X + \boxed{s(s(Y))}^{\uparrow} \Rightarrow \boxed{s(\boxed{s(X)}^{\downarrow} + Y)}^{\uparrow}.$$

We delay the discussion of how the search-control issues raised here are dealt with until Section 3.8.

3.4.4 Patch: lemma calculation

Our second technique is called *lemma calculation* and is essentially a variant of the weak fertilization strategy introduced in Section 2.3. To illustrate, consider the problem

Given: $rev(rev(t)) = t$ $\hspace{2cm}$ (3.28)

Goal: $rev(\underbrace{\boxed{rev(t)} <> h :: nil}^{\uparrow}) = \boxed{h :: t}^{\uparrow}.$ $\hspace{1cm}$ (3.29)

$$\underbrace{\hspace{3cm}}_{\textbf{blocked}}$$

Note that while the left-hand side is blocked, the right-hand side is fully rippled with respect to the equality. As a consequence, (3.28) can be used to rewrite the wave-hole on the right-hand side of (3.29). Following the weak fertilization strategy, this gives us the new goal

$$rev(rev(t) <> h :: nil) = h :: rev(rev(t)).$$

Generalizing this goal by replacing $rev(t)$ by a new, universally quantified, variable gives rise to the lemma from which the missing wave-rule is derived, i.e.

$$rev(\boxed{X} <> \boxed{(Y :: nil)}^{\uparrow}) \Rightarrow \boxed{Y :: \boxed{rev(X)}}^{\uparrow}.$$

This wave-rule unblocks (3.29) and leads to a successful proof. In lemma calculation, the generation of the new goal and subsequent generalization is performed as a separate proof attempt. The advantage of this decoupling is the ease with which lemmas can be reused. This is not the case with weak fertilization where the same lemma may be re-discovered during the same proof attempt. Note that the *half* example (presented in Section 2.3.3) is another example of where lemma calculation is applicable.

3.5 Generalizing conjectures

There are many forms a generalization step may take. For a detailed discussion of the related literature the interested reader is directed to Hummel (1987) and Hesketh (1991). Here we focus upon one particular form of generalization that involves the introduction of new, universally quantified variables.

3.5.1 Failure analysis

In order to illustrate the idea, consider the following conjecture concerning list reversal,

$$\forall t{:}list(\tau). \; rev(t) = qrev(t, nil), \tag{3.30}$$

where *qrev* is a tail recursive list reversal function (defined in Appendix 2). Note that this conjecture is a special case of the conjecture in Section 2.2.1.

The proof of (3.30) is by a 1-step induction on the list t. The associated step case gives rise to the following rippling problem.

Given:	$rev(t) = qrev(t, nil)$	(3.31)
Goal:	$rev(\boxed{h :: t}^{\uparrow}) = qrev(\boxed{h :: t}^{\uparrow}, nil)$	(3.32)

Using a wave-rule derived from the recursive definition of *rev*,

$$rev(\boxed{X :: Y}^{\uparrow}) \Rightarrow \boxed{rev(Y) <> (X :: nil)}^{\uparrow}, \qquad (3.33)$$

the goal (3.32) ripples to

$$\boxed{rev(t) <> (h :: nil)}^{\uparrow} = qrev(\boxed{h :: t}^{\uparrow}, nil). \qquad (3.34)$$

Now consider the wave-rule

$$qrev(\boxed{X :: Y}^{\uparrow}, Z) \Rightarrow qrev(Y, \boxed{X :: Z}^{\downarrow}), \qquad (3.35)$$

which comes from the recursive definition of *qrev*. Note that, while the left-hand side of (3.35) matches the right-hand side of (3.34) (i.e. $qrev(\boxed{h :: t}^{\uparrow}, nil)$), it does not lead to a proof. The resulting ripple is

$$\boxed{rev(t) <> (h :: nil)}^{\uparrow} = qrev(t, \underbrace{\boxed{h :: nil}^{\downarrow}}_{\text{blocked}}),$$

and the blockage occurs because the wave-front on the right-hand side is directed downward onto the constant *nil*, which is neither a sink nor an outwards wave-front. Any attempt to achieve a match with (3.31) is therefore doomed, since $h :: nil$ and *nil* are distinct terms. The observant reader may have noted, however, that the application of wave-rule (3.35) is actually ruled-out by the fourth precondition (see Section 3.2) of rippling:

(iv) If this wave application is prior to fertilization then any new inwards wave-front should have a sink or an outwards wave-front in its wave-hole.

This is clearly not the case in (3.34), where rippling would yield

$$\ldots = qrev(t, \underbrace{\boxed{h :: nil}^{\downarrow}}_{\text{no sink}}).$$

Note that although precondition (iv) fails, preconditions (i) to (iii) of rippling are satisfied (see Section 2.2.3, defined in Section 3.2).

3.5.2 Patch: sink speculation

Based upon our expectation of how an inductive proof involving sideways rippling should be completed, we can attempt to patch the proof attempt accordingly. In the case of the failure of precondition (iv), we are looking to introduce a fresh, universally quantified variable, e.g. a sink that will absorb the

$\boxed{h :: \ldots}^{\downarrow}$ wave-front. The problem is that we do not know how this new universally quantified variable will relate to the term structure within the original conjecture. Here, we can again use middle-out reasoning to adopt a least-commitment strategy with respect to the placement of the new, universally quantified variables within the conjecture.

Assuming the new variable is called l, we begin by postulating the schematic conjecture

$$\forall t \colon list(\tau).\forall l \colon list(\tau).\ G_1(rev(t), l) = qrev(t, F_1(l)). \tag{3.36}$$

Here, F_1 and G_1 are second-order meta-variables. Note that we have introduced two occurrences of the sink l. Each is embedded within a meta-term. The position of the meta-term $F_1(l)$ corresponds to the position within the goal that will make precondition (iv) succeed. (Discussion of the construction and positioning of the meta-term $G_1(rev(t), l)$ is delayed until Section 3.5.3.)

The proof of (3.36) is again by a 1-step induction on the list t. The resulting rippling problem now takes the form

Given: $G_1(rev(t), L) = qrev(t, F_1(L))$

Goal: $G_1(rev(\boxed{h :: t}^{\uparrow}), \lfloor l \rfloor) = qrev(\boxed{h :: t}^{\uparrow}, F_1(\lfloor l \rfloor)). \tag{3.37}$

Using wave-rule (3.35), the right-hand side of (3.37) ripples to

$$\ldots = qrev(t, \boxed{h :: F_1(\lfloor l \rfloor)}^{\downarrow}). \tag{3.38}$$

The wave-front on the right-hand side can now be absorbed by the sink l, if F_1 is instantiated to be the identity function, i.e. $\lambda x.x$. This eager fertilization step completes the proof on the right-hand side and yields

$$\ldots = qrev(t, \left\lfloor \boxed{h :: l}^{\downarrow} \right\rfloor).$$

To complete the step case, we must successfully ripple on the left-hand side, i.e.

$$G_1(rev(\boxed{h :: t}^{\uparrow}), \lfloor l \rfloor) = \ldots .$$

By wave-rule (3.33) we obtain

$$G_1(\boxed{rev(t)} <> h :: nil^{\uparrow}, \lfloor l \rfloor) = \ldots . \tag{3.39}$$

Rippling now constrains us to consider rules that manipulate outward-directed wave-fronts of the form $\ldots <> (h :: nil)$. If we assume only the definitions of *rev*, <> and the associativity of <> (i.e. (3.3)), then rippling constrains us to consider just the following three wave-rules:

$$X :: \boxed{Y <> Z}^{\uparrow} \Rightarrow \boxed{X :: Y <> Z}^{\uparrow}$$

$$X <> (\boxed{Y <> Z}^{\uparrow}) \Rightarrow \boxed{(X <> Y) <> Z}^{\uparrow}$$

$$(\boxed{X <> Y}^{\uparrow}) <> Z \Rightarrow X <> (\boxed{Y <> Z}^{\downarrow}). \qquad (3.40)$$

In the current example, it is wave-rule (3.40) that gives rise to a proof as (3.39) ripples to

$$rev(t) <> \boxed{h :: nil <> \boxed{G_2(\boxed{rev(t)} <> h :: nil}^{\uparrow}, \lfloor l \rfloor)}^{\downarrow} = \ldots .$$

Note that the application of (3.40) instantiates G_1 to be $\lambda x . \lambda y . x <> G_2(x, y)$. Finally, eager fertilization completes the proof on the left-hand side, instantiating G_2 to be a projection onto its second argument. This yields

$$rev(t) <> \left\lfloor \boxed{h :: nil <> l}^{\downarrow} \right\rfloor = \ldots .$$

Simplifying the sink on the left-hand side gives

$$rev(t) <> \left\lfloor \boxed{h :: l}^{\downarrow} \right\rfloor = qrev(t, \left\lfloor \boxed{h :: l}^{\downarrow} \right\rfloor).$$

Hence, the overall effect of this incremental process is to instantiate (3.36) to

$$\forall t : list(\tau) . \forall l : list(\tau) . \ rev(t) <> l = qrev(t, l). \qquad (3.41)$$

As a final step, we must show that (3.41) is indeed a generalization of (3.30). This means proving the conjecture

$$(\forall t : list(\tau) . \forall l : list(\tau) .$$

$$(rev(t) <> l = qrev(t, l))) \rightarrow (\forall t : list(\tau) . (rev(t) = qrev(t, nil))).$$

But specializing l to be *nil* and simplifying the antecedent gives rise to a trivial tautology.

3.5.3 Alternative generalizations

We now outline some of the alternatives that need to be considered when searching for an inductive generalization. Rippling imposes structure upon the search space of proofs in that different wave-rules suggest different generalizations. The available definitions and lemmas directly affect the generalizations that can be suggested through the analysis of failed rippling proofs.

The effect of an additional lemma

To illustrate, consider a speculative generalization of the form

$$\forall t{:}list(\tau).\forall l{:}list(\tau).\ rev(rev(t) <> F_1(l)) = G_1(t, l), \qquad (3.42)$$

where F_1 and G_1 again are second-order meta-variables. Relying upon the wave-rules derived from the definitions of *rev* and $<>$, together with the wave-rules provided by lemma (3.4), rippling instantiates (3.42) to

$$\forall t{:}list(\tau).\forall l{:}list(\tau).\ rev(rev(t) <> l) = rev(l) <> t.$$

However, now consider adding the lemma

$$\forall y{:}\tau.\forall x{:}list(\tau).\ rev(x <> (y :: nil)) = y :: rev(x).$$

This lemma gives rise to a number of wave-rules, one of which is

$$\boxed{Y :: rev(X)}^{\downarrow} \Rightarrow rev(\boxed{X <> (Y :: nil)}^{\downarrow}). \qquad (3.43)$$

In the preceding instantiation for (3.42), the rippling proof was completed by instantiating F_1 to be the identity function. With the introduction of (3.43), further rippling is possible, i.e.

$$rev(rev(t) <> rev(\boxed{F_2(\lfloor l \rfloor) <> h :: nil}^{\downarrow})) = \dots .$$

Here, eager fertilization terminates rippling by instantiating F_2 to be $\lambda x.x$, giving

$$rev(rev(t) <> rev(\left\lfloor \boxed{l <> (h :: nil)}^{\downarrow} \right\rfloor)) = \dots .$$

Turning to the right-hand side, rippling based upon (3.4) gives

$$\dots = \boxed{G_2(t, \lfloor l \rfloor) <> (h :: nil)}^{\downarrow} <> t.$$

Eager fertilization now instantiates G_2 to be a projection onto its second argument, resulting in

$$\dots = \left\lfloor \boxed{l <> (h :: nil)}^{\downarrow} \right\rfloor <> t,$$

and completing the proof of the step case. The generalization resulting from this alternative proof is

$$\forall t:list(\tau).\forall l:list(\tau).\ rev(rev(t) <> rev(l)) = l <> t.$$

The effect of repositioning a meta-variable

In the same way in which we considered alternative positions for potential wave-fronts on the right-hand side of speculative wave-rules, we also consider the positioning of meta-terms within the speculation of an inductive generalization. As promised, we return to the construction of $G_1(t, l)$ within speculation (3.36). We make a distinction between the wave-fronts that caused the failure of a sideways ripple and those that did not. We refer to them as *primary* and *secondary* wave-fronts, respectively. This terminology allows us to classify the blockage in the list reversal example (3.30) as

$$\underbrace{\boxed{rev(t)} <> (h :: nil)}_{\text{secondary}}^{\uparrow} = qrev(\underbrace{h :: t}^{\uparrow}, nil).$$
$$\text{primary}$$

For all primary wave-fronts, we introduce a set of *primary sink terms*, one for each of the blocked sideways ripples. Each primary sink term contains a new universally quantified variable. For each secondary wave-front we eagerly attempt to apply a sideways ripple by introducing occurrences of the variables associated with the primary sink terms. These occurrences are specified using meta-terms called *secondary sink terms*. Relating these ideas back to the proof patch associated with the list reversal example, we get

$$G_1(\underbrace{\boxed{rev(t)} <> (h :: nil)}_{\substack{\text{secondary} \\ \text{sink term}}}^{\uparrow}, \lfloor l \rfloor) = qrev(\underbrace{h :: t}^{\uparrow}, \underset{\substack{\text{primary} \\ \text{sink term}}}{F_1(\lfloor l \rfloor)}).$$

While the positioning of the primary sink term is fixed by the failure to apply a sideways wave-rule, the positioning of secondary sink terms will, in general, be less constrained. To illustrate, consider the left-hand side of the speculative generalization given above. If we consider the start of the rippling proof, then we get an alternative speculation of the form

$$rev(G_1(\underbrace{h :: t}^{\uparrow}, \lfloor l \rfloor)) = qrev(\underbrace{h :: t}^{\uparrow}, \underset{\substack{\text{primary} \\ \text{sink term}}}{F_1(\lfloor l \rfloor)}).$$
$$\underset{\substack{\text{secondary} \\ \text{sink term}}}{}$$

Note that this alternative speculation gives rise to the alternative generalization

$$\forall t:list(\tau).\forall l:list(\tau).\ rev(qrev(l, t)) = qrev(t, l).$$

So again we see that while the constraints of rippling provide guidance, we are not eliminating all search. However, many of the search paths will succeed, yielding alternative proofs. We return to the issues of search raised here in Section 3.8.

3.6 Case analysis

We complete our systematic analysis of precondition failures by considering the third precondition:

(iii) If this wave-rule is conditional then its condition must be provable.

The failure of this precondition suggests the need for a casesplit. Below we explore how rippling can be used to guide the application of casesplitting.

3.6.1 Failure analysis

Consider again the following goal, which arose in Section 2.5,

$$x \in \boxed{h :: (t <> l)}^{\uparrow} \leftrightarrow x \in \boxed{h :: t}^{\uparrow} \vee x \in l, \qquad (3.44)$$

where $\in$ is defined by the following conditional formulas:

$$X \in nil \leftrightarrow false$$
$$X \neq Y \rightarrow X \in Y :: Z \leftrightarrow X \in Z$$
$$X = Y \rightarrow X \in Y :: Z \leftrightarrow true.$$

By inspecting the wave-rules that arise from this definition, in particular

$$X \neq Y \rightarrow X \in \boxed{Y :: Z}^{\uparrow} \Rightarrow X \in Z, \qquad (3.45)$$

it is clear that the condition $x \neq h$ must hold in order that wave-rule (3.45) can be used to complete the rippling of (3.44). The absence of this condition corresponds to the failure of precondition (iii).

3.6.2 Patch: casesplit calculation

The suggested patch is to perform a casesplit based upon the disjunction

$$(x \neq h) \vee (x = h).$$

The disjuncts come from the condition attached to the complementary rewrite rule (see Section 2.5):

$$X = Y \rightarrow X \in \boxed{Y :: Z}^{\uparrow} \Rightarrow true. \tag{3.46}$$

Two subgoals are generated by applying the casesplit to (3.44). In the $(x \neq h)$ case, wave-rule (3.45) reduces the subgoal to

$$x \in (t <> l) \leftrightarrow x \in t \lor x \in l.$$

Strong fertilization is now applicable. In the $(x = h)$ case, where rippling is not applicable, (3.46) reduces the goal to the trivial tautology

$$true \leftrightarrow true.$$

3.7 Rotate length conjecture revisited

We now return to the rotate length conjecture and, as promised in Section 3.1, we reconsider the associated eureka steps in the light of our rippling-based proof-patching techniques. In particular, we consider both conjecture generalization and lemma discovery in turn.

3.7.1 Rotate length: conjecture generalization

Based upon the definition of *length*, rippling suggests a 1-step induction on t for conjecture (3.1). The resulting step case problem takes the form

Given: $rotate(length(t), t) = t$

Goal: $rotate(length(\boxed{h :: t}^{\uparrow}), \boxed{h :: t}^{\uparrow}) = \boxed{h :: t}^{\uparrow}.$

While the definition of *rotate* provides a matching wave-rule, i.e.

$$rotate(\boxed{s(X)}^{\uparrow}, \boxed{Y :: Z}^{\uparrow}) \Rightarrow rotate(X, \boxed{Z <> (Y :: nil)}^{\downarrow}). \tag{3.47}$$

The applicability of this wave-rule is ruled out on heuristic grounds because the second argument position of *rotate* within the goal does not contain a sink, i.e.

$$rotate(\boxed{s(length(t))}^{\uparrow}, \underbrace{\boxed{h :: t}^{\uparrow}}_{\text{no sink}}) = \boxed{h :: t}^{\uparrow}.$$

At this point, the sink speculation proof patch generates the schematic conjecture

$$\forall t{:}list(\tau).\forall l{:}list(\tau).\ rotate(length(t), F_1(t, l)) = G_1(t, l). \qquad (3.48)$$

Now consider an inductive proof of (3.48). Again rippling suggests a 1-step induction on the list t. The resulting rippling problem is

Given: $rotate(length(t), F_1(t, l)) = G_1(t, l)$

Goal: $rotate(length(\boxed{h :: t}^{\uparrow}), F_1(\boxed{h :: t}^{\uparrow}, \lfloor l \rfloor))$

$$= G_1(\boxed{h :: t}^{\uparrow}, l). \qquad (3.49)$$

The following wave-rules play a role in the instantiation of (3.49):

$$\boxed{X :: Y}^{\uparrow} <> Z \Rightarrow \boxed{X :: Y <> Z}^{\uparrow} \qquad (3.50)$$

$$\boxed{(X <> Y) <> Z}^{\downarrow} \Rightarrow X <> \boxed{Y <> Z}^{\downarrow} \qquad (3.51)$$

$$X <> \boxed{Y :: Z}^{\uparrow} \Rightarrow \boxed{X <> Y :: nil}^{\downarrow} <> Z. \qquad (3.52)$$

While (3.50) is derived from the recursive definition of $<>$, (3.51) and (3.52) come from lemmas (3.3) and (3.4), respectively. Using wave-rules (3.9), (3.50), (3.47), and (3.51), the rippling on the left-hand side is as follows.

$$rotate(length(\boxed{h :: t}^{\uparrow}), F_1(\boxed{h :: t}^{\uparrow}, \lfloor l \rfloor)) = \ldots$$

$$rotate(\boxed{s(length(t))}^{\uparrow}, F_1(\boxed{h :: t}^{\uparrow}, \lfloor l \rfloor)) = \ldots$$

$$rotate(\boxed{s(length(t))}^{\uparrow}, \boxed{h :: t <> F_2(\boxed{h :: t}^{\uparrow}, \lfloor l \rfloor)}^{\uparrow}) = \ldots$$

$$rotate(length(t), \boxed{(t <> F_2(\boxed{h :: t}^{\uparrow}, \lfloor l \rfloor)) <> h :: nil}^{\downarrow}) = \ldots$$

$$rotate(length(t), t <> (\boxed{F_2(\boxed{h :: t}^{\uparrow}, \lfloor l \rfloor) <> h :: nil}^{\downarrow})) = \ldots$$

While on the right-hand side, wave-rule (3.52) gives rise to the ripple

$$\ldots = G_1(\boxed{h :: t}^{\uparrow}, \lfloor l \rfloor)$$

$$\ldots = \boxed{G_2(\boxed{h :: t}^{\uparrow}, \lfloor l \rfloor) <> h :: nil}^{\downarrow} <> t.$$

Note that the inward directed wave-fronts on both sides of the goal are directly above sinks. Consequently, eager fertilization instantiates F_2 and G_2 to be projections onto their second arguments. This completes the step case proof, yielding

$$rotate(length(t), t <> \left\lfloor \boxed{l <> h :: nil}^{\downarrow} \right\rfloor) = \left\lfloor \boxed{l <> h :: nil}^{\downarrow} \right\rfloor <> t.$$

As a side-effect, F_1 and G_1 are respectively instantiated to

$$\lambda x.\lambda y.\ (x <> y) \text{ and } \lambda x.\lambda y.\ (y <> x).$$

In summary, the effect of the patching process is to instantiate (3.48) so that it is identical to (3.2), yielding the generalized rotate length theorem

$$\forall t{:}list(\tau).\forall l{:}list(\tau).\ rotate(length(t), t <> l) = l <> t.$$

3.7.2 Rotate length: lemma discovery

We now consider the other side of the coin: given the generalized version of the rotate length conjecture, can we construct an inductive proof without prior knowledge of lemmas (3.3) and (3.4)? In the case of (3.2), rippling suggests a 1-step list induction on t giving rise to the problem

Given: $rotate(length(t), t <> l) = l <> t$

Goal: $rotate(length(\boxed{h :: t}^{\uparrow}), \boxed{h :: t}^{\uparrow} <> \lfloor l \rfloor)$
$$= \lfloor l \rfloor <> \boxed{h :: t}^{\uparrow}.$$

The corresponding rippling proof is

$$rotate(length(\boxed{h :: t}^{\uparrow}), \boxed{h :: t}^{\uparrow} <> \lfloor l \rfloor) = \lfloor l \rfloor <> \boxed{h :: t}^{\uparrow}$$

$$rotate(\boxed{s(length(t))}^{\uparrow}, \boxed{h :: t}^{\uparrow} <> \lfloor l \rfloor) = \lfloor l \rfloor <> \boxed{h :: t}^{\uparrow}$$

$$rotate(\boxed{s(length(t))}^{\uparrow}, \boxed{h :: t <> \lfloor l \rfloor}^{\uparrow}) = \lfloor l \rfloor <> \boxed{h :: t}^{\uparrow}$$

$$rotate(length(t), \underbrace{\boxed{t <> \lfloor l \rfloor <> h :: nil}^{\downarrow}}_{\textbf{blocked}}) = \underbrace{\lfloor l \rfloor <> \boxed{h :: t}^{\uparrow}}_{\textbf{blocked}}.$$

As no more wave-rules are applicable, lemma speculation is triggered. Based upon the above blockage terms, the following two wave-rule speculations are generated:

$$X <> Y <> Z :: nil^{\downarrow} \Rightarrow X <> \boxed{F_1(\underline{Y}, X, Z)}^{\downarrow} \tag{3.53}$$

$$X <> Y :: Z^{\uparrow} \Rightarrow \boxed{G_1(\underline{X}, Y, Z)}^{\downarrow} <> Z. \tag{3.54}$$

While (3.53) targets the left-hand side of the blocked goal, (3.54) targets the right-hand side. The application of these speculative wave-rules unblocks the goal yielding

$$rotate(length(t), t <> \left\lfloor \boxed{F_1(\underline{l}, h, t)}^{\downarrow} \right\rfloor) = \left\lfloor \boxed{G_1(\underline{l}, h, t)}^{\downarrow} \right\rfloor <> t.$$

Although both sides of the equality are fully rippled, the identity of F_1 and G_1 remain unknown. Their identity is determined by considering the constraints imposed by (3.53) and (3.54): in particular, (3.53) when X is *nil* and (3.54) when Z is *nil*. In addition, strong fertilization also imposes constraints, i.e. multiple instances of the same sink must be instantiated identically. The resulting instantiations for F_1 and G_1 respectively are

$$\lambda x.\lambda y.\lambda z. \ (x <> (y :: nil)) \text{ and } \lambda x.\lambda y.\lambda z. \ (x <> (y :: nil)).$$

Applying these instantiations to (3.53) and (3.54) gives rise to the wave-rules

$$X <> Y <> Z :: nil^{\downarrow} \Rightarrow X <> \boxed{Y <> (Z :: nil)}^{\downarrow}$$

$$X <> Y :: Z^{\uparrow} \Rightarrow \boxed{X <> (Y :: nil)}^{\downarrow} <> Z.$$

Note that these wave-rules correspond to the missing lemmas, i.e. (3.3) and (3.4). Once discovered, the lemmas must be proved in order to ensure the soundness of the main proof.

3.7.3 An automated reasoning challenge

The observant reader may have noticed that the lemmas that guided the generalization of the rotate length conjecture are the same lemmas that are discovered during the proof of the generalized conjecture. Discovering the lemmas and the generalization simultaneously is beyond the current state-of-the-art for automated reasoning systems. However, the reader should note that the lemma discovery and generalization techniques can work in concert if the missing lemmas are not key to the generalization.

3.8 Implementation and results

The work discussed in this chapter has been implemented within an extension to the *C LAM* proof planner (Bundy *et al.*, 1990b). The extension makes use of the higher-order features of λ-Prolog (Nadathur & Miller, 1988). As has been indicated throughout this chapter, we use the requirement of skeleton preservation and the wave annotations to constrain the higher-order unification problems we pass to λ-Prolog. An alternative to this pragmatic approach is presented in (Hutter & Kohlhase, 1997), where essentially the structure preservation constraints of rippling are embedded within the unification algorithm. While this approach has not been applied to the problem of generalization, we believe that there are no technical reasons why it should not work. As indicated in Sections 3.4 and 3.5, the processes of patch specification and construction both involve search. We manage this search through the use of an iterative deepening search strategy to enable alternative branches within the search space to be explored. Moreover, we employ a simple counter-example finder to filter the candidate instantiations of the schematic conjecture. The proof planner is used recursively to plan all conjectures.

Our test results are presented in Tables 3.1, 3.2, and 3.3. While Table 3.1 presents the conjectures, Tables 3.2 and 3.3 provide the lemmas and generalizations, respectively. Note that Table 3.2 contains both the lemmas that are automatically discovered by the system as well as those used in the generalizing conjectures.

The proofs of all the example conjectures[1] given in Table 3.1 are discovered completely automatically. These proofs are based only upon definitions supplied by the user. Except for the generalization examples, all additional lemmas are discovered automatically by the system.

The example conjectures in Table 3.1 are classified under the four critics. In the case of lemma discovery, conjectures $T1$–$T13$, $T22$–$T26$ and $T48$–$T50$ required only the relatively weak strategy of lemma calculation. Examples $T14$–$T21$ required lemma speculation while $T27$–$T35$ required generalization. Note that different generalizations were obtained depending upon the available lemmas. All the examples that required induction revision, lemma speculation or generalization fall into a class of inductive theorem that are not uncommon but are acutely difficult to prove. Finally, examples $T22$–$T26$ and $T48$–$T50$ illustrate the need for multiple critics in patching some conjectures.

---

[1] The examples come from a number of sources that include Aubin (1975), Boyer and Moore (1979), Manna and Waldinger (1985), and Walsh (1994).

Table 3.1 *Example conjectures.*

The numbered columns denote (i) induction revision (1-step $\mapsto$ 2-step), (ii) lemma discovery, (iii) generalization and (iv) casesplit. Moreover, the references in the columns (ii) and (iii) refer to Tables 3.2 and 3.3, respectively. Note that $nth(X, Y)$ denotes the list constructed by removing the first Xth elements from Y. Note also that *fac*, *exp* and $\times$ denote factorial, exponentiation and multiplication, while *qfac*, *qexp* and *mult* denote tail recursive versions, respectively.

No.	Conjecture	(i)	(ii)	(iii)	(iv)
$T1$	$double(X) = X + X$		$L1$		
$T2$	$length(X <> Y) = length(Y <> X)$		$L2$		
$T3$	$length(X <> Y) = length(Y) + length(X)$		$L1$		
$T4$	$length(X <> X) = double(length(X))$		$L2$		
$T5$	$length(rev(X)) = length(X)$		$L3$		
$T6$	$length(rev(X <> Y))$ $= length(X) + length(Y)$		$L2$		
$T7$	$length(qrev(X, Y)) = length(X) + length(Y)$		$L1$		
$T8$	$nth(X, nth(Y, Z)) = nth(Y, nth(X, Z))$		$L4/5$		
$T9$	$nth(W, nth(X, nth(Y, Z)))$ $= nth(Y, nth(X, nth(W, Z)))$		$L6/7$		
$T10$	$rev(rev(X)) = X$		$L8$		
$T11$	$rev(rev(X) <> rev(Y)) = Y <> X$		$L9/10$		
$T12$	$qrev(X, Y) = rev(X) <> Y$		$L11$		
$T13$	$half(X + X) = X$		$L1$		
$T14$	$ordered(isort(X))$		$L12$		
$T15$	$X + s(X) = s(X + X)$		$L1$		
$T16$	$even(X + X)$		$L1$		
$T17$	$rev(rev(X <> Y))$ $= rev(rev(X)) <> rev(rev(Y))$		$L8$		
$T18$	$rev(rev(X) <> Y) = rev(Y) <> X$		$L11/13$		
$T19$	$rev(rev(X)) <> Y = rev(rev(X <> Y))$		$L8$		
$T20$	$even(length(X <> X))$		$L2$		
$T21$	$rotate(length(X), X <> Y) = Y <> X$		$L11/13$		
$T22$	$even(length(X <> Y))$ $\leftrightarrow even(length(Y <> X))$	$\star$	$L14$		
$T23$	$half(length(X <> Y))$ $= half(length(Y <> X))$	$\star$	$L15$		
$T24$	$even(X + Y) \leftrightarrow even(Y + X)$	$\star$	$L16$		
$T25$	$even(length(X <> Y))$ $\leftrightarrow even(length(Y) + length(X))$	$\star$	$L16$		
$T26$	$half(X + Y) = half(Y + X)$	$\star$	$L17$		
$T27$	$rev(X) = qrev(X, nil)$			$G1$	
$T28$	$revflat(X) = qrevflat(X, nil)$			$G2$	
$T29$	$rev(qrev(X, nil)) = X$			$G3/4$	
$T30$	$rev(rev(X) <> nil) = X$			$G5/6$	
$T31$	$qrev(qrev(X, nil), nil) = X$			$G7/8$	
$T32$	$rotate(length(X), X) = X$			$G9$	

(Cont.)

Table 3.1 *(Cont.)*

No.	Conjecture	(i)	(ii)	(iii)	(iv)
$T33$	$fac(X) = qfac(X, 1)$			$G10$	
$T34$	$X * Y = mult(X, Y, 0)$			$G11$	
$T35$	$exp(X, Y) = qexp(X, Y, 1)$			$G12$	
$T36$	$X \in Y \rightarrow X \in (Y <> Z)$				$\star$
$T37$	$X \in Z \rightarrow X \in (Y <> Z)$				$\star$
$T38$	$(X \in Y) \vee (X \in Z) \rightarrow X \in (Y <> Z)$				$\star$
$T39$	$X \in nth(Y, Z) \rightarrow X \in Z$				$\star$
$T40$	$X \subset Y \rightarrow (X \cup Y = Y)$				$\star$
$T41$	$X \subset Y \rightarrow (X \cap Y = X)$				$\star$
$T42$	$X \in Y \rightarrow X \in (Y \cup Z)$				$\star$
$T43$	$X \in Y \rightarrow X \in (Z \cup Y)$				$\star$
$T44$	$(X \in Y) \wedge (X \in Z) \rightarrow (X \in Y \cap Z)$				$\star$
$T45$	$X \in insert(X, Y)$				$\star$
$T46$	$X = Y \rightarrow (X \in insert(Y, Z) \leftrightarrow true)$				$\star$
$T47$	$X \neq Y \rightarrow (X \in insert(Y, Z) \leftrightarrow X \in Z)$				$\star$
$T48$	$length(isort(X)) = length(X)$		$L18$		$\star$
$T49$	$X \in isort(Y) \rightarrow X \in Y$		$L19$		$\star$
$T50$	$count(X, isort(Y)) = count(X, Y)$		$L20/21$		$\star$

Table 3.2 *Lemmas.*

No.	Lemma
$L1$	$X + s(Y) = s(X + Y)$
$L2$	$length(X <> Y :: Z) = s(length(X <> Z))$
$L3$	$length(X <> Y :: nil) = s(length(X))$
$L4$	$nth(s(W), nth(X, Y :: Z)) = nth(W, nth(X, Z))$
$L5$	$nth(s(V), nth(s(W), X :: Y :: Z)) = nth(s(V), nth(W, X :: Z))$
$L6$	$nth(s(V), nth(W, nth(X, Y :: Z))) = nth(V, nth(W, nth(X, Z)))$
$L7$	$nth(s(U), nth(V, nth(s(W), X :: Y :: Z)))$ $= nth(s(U), nth(V, nth(W, X :: Z)))$
$L8$	$rev(X <> (Y :: nil)) = Y :: rev(X)$
$L9$	$rev(X <> (Y <> Z :: nil)) = Z :: rev(X <> Y)$
$L10$	$rev((X <> Y :: nil) <> nil) = Y :: rev(X <> nil)$
$L11$	$(X <> (Y :: nil)) <> Z = X <> (Y :: Z)$
$L12$	$ordered(Y) \rightarrow ordered(insert(X, Y))$
$L13$	$(X <> Y) <> Z :: nil = X <> (Y <> Z :: nil)$
$L14$	$even(length(W <> Z)) \leftrightarrow even(length(W <> X :: Y :: Z))$
$L15$	$length(W <> X :: Y :: Z) = s(s(length(W <> Z)))$
$L16$	$even(X + Y) \leftrightarrow even(X + s(s(Y)))$
$L17$	$X + s(s(Y)) = s(s(X + Y))$
$L18$	$length(insert(X, Y)) = s(length(Y))$
$L19$	$X \neq Y \rightarrow (X \in insert(Y, Z) \rightarrow X \in Z)$
$L20$	$count(X, insert(X, Y)) = s(count(X, Y))$
$L21$	$X \neq Y \rightarrow (count(X, insert(Y, Z)) = count(X, Z))$
$L22$	$(X <> Y) <> Z = X <> (Y <> Z)$
$L23$	$(X * Y) * Z = X * (Y * Z)$
$L24$	$(X + Y) + Z = X + (Y + Z)$

Table 3.3 *Generalized conjectures.*

The lemmas used to motivate each generalization are indicated in the right-hand column.

No.	Generalization	Lemmas
$G1$	$rev(X) <> Y = qrev(X, Y)$	$L22$
$G2$	$revflat(X) <> Y = qrevflat(X, Y)$	$L22$
$G3$	$rev(qrev(X, Y)) = rev(Y) <> X$	$L11$
$G4$	$rev(qrev(X, rev(Y))) = Y <> X$	$L8\&L11$
$G5$	$rev(rev(X) <> Y) = rev(Y) <> X$	$L11$
$G6$	$rev(rev(X) <> rev(Y)) = Y <> X$	$L8\&L11$
$G7$	$qrev(qrev(X, Y), nil) = rev(Y) <> X$	$L11$
$G8$	$qrev(qrev(X, rev(Y)), nil) = Y <> X$	$L8\&L11$
$G9$	$rotate(length(X), X <> Y) = Y <> X$	$L11\&L22$
$G10$	$fac(X) * Y = qfac(X, Y)$	$L23$
$G11$	$(X * Y) + Z = mult(X, Y, Z)$	$L24$
$G12$	$exp(X, Y) * Z = qexp(X, Y, Z)$	$L23$

Table 3.4 *Precondition failures and patches for rippling.*

The association between precondition failure and patches for the ripple method are shown. Note that $\star$, o and • denote success, partial success and failure, respectively.

Precondition	Generalization	Case analyse	Induction revision	Lemma discovery
1	$\star$	$\star$	$\star$	$\star$
2	$\star$	$\star$	o	•
3	$\star$	•		
4	•			

3.9 Summary

In this chapter we have described how the constraints of rippling can be used productively to overcome failed proof attempts. In particular, we have shown how missing lemmas can be discovered, conjectures can be generalized, induction-rule selection can be revised, and case analyses may be suggested.

The association between precondition failures and patches to the ripple method is summarized in Table 3.4. The structure of this table reflects the systematic nature of our analysis. Moreover, it underlies the effectiveness of rippling as a search control technique.

4

A formal account of rippling

As explained in Section 1.1.4, rippling is a heuristic that reflects a common pattern of reasoning found in theorem-proving: one wants to prove a goal using a given, and rewriting is used to transform the goal to the point where the given can be used. As noted in Chapter 2, there are various complications, such as multiple goals and givens and universally quantified givens with corresponding sinks. However, the general pattern is the same and can be codified by methods used in proof-planning.

To mechanize this common pattern of reasoning, rippling generalizes rewriting, so that semantic information is used to guide proof construction. The user has expectations (encoded by the proof-plan methods) about how the proof should proceed, namely that differences between the goal and the givens should be minimized. Annotations provide a kind of abstraction that is used to minimize these differences. Under this abstraction, the identity of the different symbols is ignored and one just differentiates whether they belong to the skeleton or not. Rippling constitutes an extension of rewriting that uses these annotations to drive proofs forward in a goal-directed way. Differences are monotonically decreased and rippling terminates with success or failure depending on whether the givens can be used or not.

In this chapter we consider how the concepts described above can be formalized. There is no one best formalization, so we keep our discussion abstract, when possible, and first discuss what properties are desired. Afterwards we consider different formalizations and implementations of rippling with these properties.

4.1 General preliminaries

Before going into details particular to rippling, we need a few general preliminaries concerning terms and rewriting. We use standard definitions here (see, for example, Baader and Nipkow (1998)).

82

4.1.1 Terms and positions

Terms are built from function symbols and variables in the usual way. To make clear which function symbols are available in a given context, and what arity they have, we work with signatures. A *signature* Σ is a set of function symbols where each $f \in \Sigma$ is associated with a non-negative integer n, the *arity* of f. We will write $\Sigma^{(n)}$ to denote the set of function symbols in Σ of arity n. Functions in $\Sigma^{(0)}$ are constants. The set of terms $\mathcal{T}_{\Sigma(\mathcal{X})}$ built from Σ and a set of variables $\mathcal{X}$ (disjoint from Σ) is inductively defined: $\mathcal{X} \subseteq \mathcal{T}_{\Sigma(\mathcal{X})}$ and if $f \in \Sigma^{(n)}$ and $t_1, \ldots, t_n \in \mathcal{T}_{\Sigma(\mathcal{X})}$, then $f(t_1, \ldots, t_n) \in \mathcal{T}_{\Sigma(\mathcal{X})}$. We write $\mathcal{T}_{\Sigma}$ for ground terms ($\mathcal{X} = \emptyset$) and omit both Σ and $\mathcal{X}$ and simply write $\mathcal{T}$ when these are implicitly given or their identity is unimportant.

Note that, in many applications, it is helpful to impose a *sort discipline* where terms and operations on them are partitioned into different classes. For instance, in the examples in Chapter 2, the reader will see that terms include both the terms and formulas in a first-order theory. Here we might use two sorts to distinguish these two kinds of entities from each other. However, in order to simplify notation, we will not consider sorted extensions in this book, although the formalizations of rippling we give can easily be so generalized.

As we have seen, terms may be visualized as trees. It will be useful to refer to positions in terms and have notation for term replacement. To do this, we associate a path address, represented by a string of positive integers, with each node in the tree. Formally, we define $Pos(s)$, the set of positions in the term s, by

$$Pos(X) = \{\epsilon\}$$

where $X \in \mathcal{X}$

$$Pos(f(s_1, \ldots, s_n)) = \{\epsilon\} \cup \bigcup_{i=1}^{n} \{ip \mid p \in Pos(s_i)\}.$$

The symbol ϵ denotes the empty string.

Positions can be partially ordered by

$$p \leq q \equiv \exists p'. \, pp' = q.$$

The position p is *above* q if $p \leq q$, and p is *strictly above* q if $p < q$ (i.e. $p \leq q$ and $p \neq q$); *below* is defined analogously.

For $p \in Pos(s)$, the *subterm of a term s at position p*, written s/p, is defined by

$$s/\epsilon = s$$
$$f(s_1, \ldots, s_n)/ip = s_i/p.$$

For $p \in Pos(s)$, we write $s[t]_p$ to denote the term obtained from s by replacing the subterm at position p by t as follows:

$$s[t]_\epsilon = t$$
$$f(s_1, \ldots, s_n)[t]_{ip} = f(s_1, \ldots, s_i[t]_p, \ldots s_n).$$

Sometimes we will talk about some distinguished occurrence of a subterm t of s, which means that there is some position p where $s/p = t$. We often write this as $s[t]$, leaving p implicit, and write $s[t']$ to denote replacing this occurrence by t', i.e. $s[t']_p$.

Finally, we write $Vars(s)$ to denote the set of variables occurring in s, i.e.

$$Vars(s) = \{X \in \mathcal{X} \mid \exists p \in Pos(s). s/p = X\}.$$

For example, let t be the term $f(g(X), h(a, X, Y))$. Then $Pos(t)$ is the set $\{\epsilon, 1, 11, 2, 21, 22, 23\}$. The position 1 is strictly above 11, but unrelated to 21. The subterm of t at position 21, i.e. $t/21$ is a, $t[g(b)]_{21}$ denotes the replacement of this term by $g(b)$ and is $f(g(X), h(g(b), X, Y))$. Finally, $Vars(t) = \{X, Y\}$.

4.1.2 Substitution and rewriting

A $(\mathcal{T}_{\Sigma(\mathcal{X})}-)$ *substitution* is a function $\sigma : \mathcal{X} \to \mathcal{T}_{\Sigma(\mathcal{X})}$ such that $\sigma(X) \neq X$ for only finitely many $X \in \mathcal{X}$. We define the domain of σ, $Dom(\sigma)$ by $\{X \mid \sigma(X) \neq X\}$ and may write σ as the set of pairs $\{t_1/X_1, \ldots, t_n/X_n\}$, where $X_i \in Dom(\sigma)$ and $t_i = \sigma(X_i)$ for all $1 \leq i \leq n$. Any substitution σ can be extended to a mapping $\hat{\sigma} : \mathcal{T}_{\Sigma(\mathcal{X})} \to \mathcal{T}_{\Sigma(\mathcal{X})}$ in the standard way: $\hat{\sigma}(X) = \sigma(X)$ for all $X \in \mathcal{X}$, and $\hat{\sigma}(f(t_1, \ldots, t_n)) = f(\hat{\sigma}(t_1), \ldots, \hat{\sigma}(t_n))$, for each $f \in \Sigma^{(n)}$. We will subsequently identify the substitution σ with its extension $\hat{\sigma}$. Note that there is a simple relationship between substitution and term replacement: applying the substitution σ to s is equivalent to simultaneously replacing in s each occurrence of each X in $Vars(s)$ with $\sigma(X)$.

We now define several different kinds of relations and orders. Let R be a binary relation, $R \subseteq \mathcal{T}_{\Sigma(\mathcal{X})} \times \mathcal{T}_{\Sigma(\mathcal{X})}$. R is *compatible with Σ-contexts* if $s \; R \; s'$ implies $t[s]_p \; R \; t[s']_p$ for all terms t and all positions $p \in Pos(t)$. Note that here, and elsewhere, we often write binary relations (and functions) using infix notation. R is *closed under substitutions* if whenever $s \; R \; t$ then $\sigma(s) \; R \; \sigma(t)$, for every substitution σ and $s, t \in \mathcal{T}_{\Sigma(\mathcal{X})}$. R is a *rewrite relation* iff it is compatible with Σ-contexts and closed under substitutions. A *strict order* is a transitive and irreflexive relation. When R is both a rewrite relation and a well-founded strict order, then R is called a *reduction order*.

Rewrite relations usually are defined indirectly via sets of rewrite rules. A *rewrite rule* is a pair of terms $\langle l, r \rangle$ in $\mathcal{T}_{\Sigma(\mathcal{X})} \times \mathcal{T}_{\Sigma(\mathcal{X})}$, where $Vars(r) \subseteq Vars(l)$. As is customary, we write such pairs by separating them with the binary symbol $\Rightarrow$, for example

$$(X + Y) + Z \Rightarrow X + (Y + Z).$$

Now, given a set of rules $\mathcal{R}$, we can define a rewrite relation $\Rightarrow_{\mathcal{R}}$, where for all $s, t \in \mathcal{T}_{\Sigma(\mathcal{X})}$, $s \Rightarrow_{\mathcal{R}} t$ iff there exists a $l \Rightarrow r \in \mathcal{R}$, $p \in Pos(s)$, and $\sigma \in \mathcal{T}_{\Sigma(\mathcal{X})}$-substitution such that $s/p = \sigma(l)$ and $t = s[\sigma(r)]_p$. This is essentially just an alternative formalization of the rewrite rule of inference given in Section 1.2, without conditions. It is easy to check that $\Rightarrow_{\mathcal{R}}$ is a rewrite relation. In practice, we will often employ certain notational shorthands and leave details implicit. For example, we say that $s[s']$ rewrites to $s[\sigma(r)]$ when it is the case that $\sigma(l) = s'$ for some subterm s' of s and some rule $l \Rightarrow r \in \mathcal{R}$. Also, we write $s[\sigma(l)] \Rightarrow_{\mathcal{R}} s[\sigma(r)]$ leaving implicit that $\sigma(l)$ matches some subterm of s. When it does not cause confusion, we may omit the subscript $\mathcal{R}$ from the rewrite relation $\Rightarrow_{\mathcal{R}}$, relying on context to distinguish the rewrite rule constructor $\Rightarrow$ from the rewrite relation it defines. It is straightforward to extend this account to include conditional rewrite rules.

Operationally, one implements rewriting using *matching*. Given a term s, one computes terms t where $s \Rightarrow_{\mathcal{R}} t$. This is done by considering in turn each subterm s' of s and attempting to match s' with the left-hand side of some rewrite rule $l \Rightarrow r \in \mathcal{R}$ and proving any condition of the rewrite rule. When this succeeds, we apply the resulting substitution, in the way described above, to give t. Since rewrite rules are often applied to terms in $\mathcal{T}_{\Sigma}$ (without variables), if the result is also to be a term in $\mathcal{T}_{\Sigma}$, then all of the variables in r must also occur in l. This accounts for the previously stated requirement that for a set of rewrite rules $\mathcal{R}$, $Vars(r) \subseteq Vars(l)$ for every rewrite rule $l \Rightarrow r \in \mathcal{R}$.

4.1.3 Notation

The following notation is useful for building sets. Given a set S, $\mathcal{P}(S)$ denotes the powerset of S, and $\mathcal{P}_1(S)$ denotes the set of non-empty subsets of S, i.e. $\mathcal{P}(S) \setminus \{\emptyset\}$. Similarly, $\mathcal{F}(S)$ [respectively $\mathcal{F}_1(S)$] denotes the set of [non-empty] finite subsets of S. As shorthand, we will write $[n]$ to represent the set $\{i \mid 1 \leq i \leq n\}$.

Finally, a warning about arrows! In order to contrast rippling with rewriting, in this chapter we will use $\Rightarrow$ only for rewriting, and introduce a separate symbol, $\Rightarrow$, for rippling. Moreover, analogous to the distinction we just made in Section 4.1.2 between $\Rightarrow$, used to define a set of rewrite rules $\mathcal{R}$, and the rewrite relation $\Rightarrow_{\mathcal{R}}$ defined by $\mathcal{R}$, we will make a similar distinction between the symbol $\Rightarrow$, used to define a set of wave-rules $\mathcal{W}$, and the rippling relation $\Rightarrow_{\mathcal{W}}$ defined by $\mathcal{W}$. In both cases, we will overload notation by dropping the subscript when the set of rules is clear from context or unimportant. In subsequent chapters we will return to our previous convention, and overload $\Rightarrow$ for rippling as well.

4.2 Properties of rippling

4.2.1 Preliminaries

To formalize the properties that we desire from rippling, without committing ourselves to a particular implementation, we first give an abstract presentation, assuming only the existence of certain sets and functions. Concrete formalizations will be given later.

Let a signature Σ, set of variables $\mathcal{X}$, a set of (*unannotated*) *terms* $\mathcal{T}$ ($= \mathcal{T}_{\Sigma(\mathcal{X})}$) over Σ and $\mathcal{X}$, and a rewrite relation $\Rightarrow\ \subseteq\ \mathcal{T} \times \mathcal{T}$ be given. Our formalization of rippling is based on a set $\mathcal{A}$ and a binary relation $\Rightarrow_{\mathcal{W}}\ \subseteq\ \mathcal{A} \times \mathcal{A}$ over this set. Members of $\mathcal{A}$ are *annotated terms* and $\Rightarrow_{\mathcal{W}}$ is the *rippling relation*. The requirements on $\mathcal{A}$ and $\Rightarrow_{\mathcal{W}}$ are weaker than for their unannotated counterparts. In particular, the elements of $\mathcal{A}$ are simply elements of a set and need not be generated from a signature (so formally we are abusing the phrase *term*, although in practice members of $\mathcal{A}$ often are terms or can be thought of as terms over a signature with additional restrictions). We also do not insist that $\Rightarrow_{\mathcal{W}}$ is a rewriting relation. The reason for these weaker requirements will become clearer in Section 4.4.2 when we commit to a particular formalization of rippling. The rough intuition, though, is that annotation provides control information for rippling by labeling parts of terms and only restricted kinds of labelings are meaningful. It turns out that the collection of meaningfully labeled terms is a set, but not a freely generated one.

To relate annotated and unannotated terms we require two functions: a function *erase* : $\mathcal{A} \rightarrow \mathcal{T}$ that computes the *erasure* of an annotated term, and a function *skel* : $\mathcal{A} \rightarrow \mathcal{P}_1(\mathcal{T})$ that computes the set of *skeletons* of an annotated term. Note that, aside from giving types, we have not said anything about what annotated terms, rippling, and the erasure and skeleton function actually are. Later we will give several different instances of these. Our goal now is instead to specify some relationships that should hold between annotated terms and rippling and their unannotated counterparts.

4.2.2 Properties of rippling

We have described rippling as extending rewriting to use semantic information to guide proof construction. A natural requirement then is that this additional information, represented as term annotations, only allows us to prove
what is provable without the annotations. Hence, given a rewrite relation $\Rightarrow_{\mathcal{R}}$
and a corresponding ripple relation $\Rightarrow_{W}$, this amounts to showing that whenever two annotated terms are in the rippling relation $\Rightarrow_{W}$, then their unannotated counterparts are in the rewriting relation $\Rightarrow_{\mathcal{R}}$. In practice this means
that rippling steps can be "simulated" by rewriting steps on their unannotated
erasures.

We have also described rippling as being goal directed and reducing the difference between the goal and the givens. Goal directedness means that rippling
aims to use the givens, and this corresponds to skeleton preservation: the image
of the givens should remain intact within the goal. In the case of an annotated
term having a single skeleton, this means that the skeleton may not change
during rippling. If there are multiple skeletons (which arises when rippling simultaneously towards multiple givens, whereby each given corresponds to a
skeleton (cf. Section 2.4)), then we shall demand something weaker: skeleton
preservation means that no new skeletons are created, although the number of
skeletons may decrease.

We also require that the rippling relation is well-founded. Combined with
skeleton preservation, this means that each rippling step makes progress in
directing the derivation towards at least one of the givens. Termination itself
has other practical benefits. If rippling does not lead to fertilization then, since
we discover this in finite time, we can backtrack and try other possibilities,
e.g. other inductions. We can also use critics (see Chapter 3) to analyze why
rippling has failed.

Summarizing, these requirements are the following.

Simulation: If s ripples to t, then *erase(s)* rewrites to *erase(t)* in the original (unannotated) theory, i.e.

$$\forall s{:}\mathcal{A}.\forall t{:}\mathcal{A}.\, s \Rightarrow_{W} t \;\rightarrow\; erase(s) \Rightarrow_{\mathcal{R}} erase(t).$$

Skeleton preservation: If s ripples to t, then the skeletons of t are a subset
of the skeletons of s, i.e.

$$\forall s{:}\mathcal{A}.\forall t{:}\mathcal{A}.\, s \Rightarrow_{W} t \;\rightarrow\; skel(t) \subseteq skel(s).$$

Termination: Rippling terminates, i.e. the relation $\Rightarrow_{W}$ is well founded.

These properties constitute minimal requirements that we will demand
of any formalization (and ultimately implementation) of rippling. The first

property alone is necessary and sufficient for rippling to be *correct* in the sense that we can only use it to prove propositions provable using $\Rightarrow_{\mathcal{R}}$. Taken together though, the properties are only necessary conditions for rippling to be *effective*; they are trivially, but uselessly, satisfied when $\Rightarrow_W$ is the empty set. However, there is a limit to how much more can usefully be formalized. Rippling is just a heuristic and any notion of effectiveness (when it is short of completeness) is itself informal; the effectiveness of any implementation must, in the end, be validated empirically. We address the empirical effectiveness of rippling in Chapter 5.

4.3 Implementing rippling: generate-and-test

In this section we present a very simple realization of rippling that has the simulation and skeleton-preservation properties and suffices to carry out the kinds of proofs considered in Chapters 1 and 2. We shall delay the issue of termination until Section 4.7. Our presentation here is somewhat artificial in that it leads to an implementation of rippling that is too inefficient for practical use. However, it illustrates the main ideas, and is related to other, more realistic, implementations.

In Section 1.3 we informally motivated annotation as a way of marking parts of terms. This marking indicates which parts of a term correspond to the givens (skeletons) and which parts do not (wave-fronts). These markings guide rippling by highlighting the parts of the term that can be transformed; so under this view, adding annotations adds semantic information that guides rippling. In this section we will take a different view where annotations serve not as a guide, but rather as a check. This leads to a simple implementation based on generate-and-test. Afterwards, we will show how this can be made more efficient, essentially by interleaving the generation with the testing.

Consider the example given in Section 1.4 where we have the given $a + b = 42$ and the goal $((c+d)+a)+b = (c+d)+42$. We can apply the rewrite rule for the associativity of plus in three different ways to this goal. We annotated the goal as

$$(\boxed{(c+d)+a}^{\uparrow}) + b = \boxed{(c+d)+42}^{\uparrow}, \tag{4.1}$$

so that its skeleton is identical to the given. Then we looked at how the results of each possible rewriting could also be annotated so that a skeleton of the given remains. All three could be so annotated, but two of the three were rejected for not making progress.

Rippling can be understood, and even implemented, as ordinary rewriting augmented with these additional tests. The test is based on checking that the skeleton remains after rewriting and that progress is made. Central to formalizing this is clarifying exactly what constitutes a skeleton. We first turn our attention to this question and then afterwards formalize annotated terms and the rippling relation.

4.3.1 Embeddings

Figure 1.2 in Section 1.1.4 presents the intuitive idea that rippling manipulates a term (the goal) in which other terms (the givens) are reflected, with possible disturbances. This corresponds to a formal notion called an embedding. An *embedding* is a relation, $\mathcal{E} \subseteq \mathcal{T} \times \mathcal{T}$; viewing terms as ordered, labeled trees, then for $s, t \in \mathcal{T}$, $s \mathcal{E} t$ holds when there is an injective map from the nodes of s to t that is label- and order-preserving. This can be formalized inductively as follows. For terms $s = f(s_1, \ldots, s_n)$ and $t = g(t_1, \ldots, t_m)$, $s \mathcal{E} t$ holds iff

(i) $\exists i. s \mathcal{E} t_i$; or
(ii) $f = g$ and $\forall i \in [n]. s_i \mathcal{E} t_i$.

Some observations are in order here. First, $\mathcal{E}$ is decidable. The above inductive definition corresponds to a recursive program: given an $s, t \in \mathcal{T}$, we decide $s \mathcal{E} t$ by recursion on t, non-deterministically choosing between cases (i) and (ii), when both apply. In both cases, the recursive call is on a proper subterm of t; hence, this program terminates in non-deterministic linear time (in the height of t) and this gives rise to a deterministic algorithm that runs in exponential time. However, since both s and t have only linearly many subterms, there are only quadratically many subproblems that can arise in deciding $s \mathcal{E} t$; thus using dynamic programming (or, equivalently, memoization to avoid repeated solutions to subproblems) gives rise to a polynomial time algorithm.

Second, when executing the above algorithm on inputs s and t, it is a simple matter to keep track of occurrences of function symbols g in t that are in the skeleton; these are simply those g that are equal to an f in case (ii). For example, consider the case when s is $a + b = 42$ and t is $((c + d) + a) + b = (c + d) + 42$. Then $s \mathcal{E} t$ holds, and if we mark the symbols in t that are *not* in the skeleton by surrounding them with wave-front markers, then t is annotated as above in (4.1). We will see that determining the existence of an embedding, and returning annotations, has many uses.

Third, given t, we can generate in exponential time all possible s that are embedded in t. To see that this is a lower bound, note that there are exponentially many s that are embeddable in the term $f_1(f_2(\ldots(f_n(a))\ldots))$, namely any sequence containing any (ordered) combination of the f_i followed by a. To see that this is an upper bound, observe that, given t, there are two possibilities for each function symbol occurrence in t that is not a leaf: either it is included in s or not depending on whether case (i) or (ii) is picked. We can recurse through t and apply each case to each function symbol to generate all the embeddings. This procedure (actually, a slight generalization of it) will be used later in Section 4.8 to generate wave-rules from rewrite rules.

Finally, as the reader may check, $\mathcal{E}$ is a rewrite relation.

4.3.2 Annotated terms and rippling

Using the notion of an embedding, we can now formalize the kind of simple-minded generate-and-test implementation described previously. We begin by concretizing $\mathcal{A}$ and $\Rightarrow_w$.

Let Σ, $\mathcal{X}$, $\mathcal{T}$, and $\Rightarrow$ be given. Now we define $\mathcal{A}$ to be $\mathcal{F}_1(\mathcal{T}) \times \mathcal{T}$; hence, each member of $\mathcal{A}$ is a pair $\langle\{s_1, \ldots, s_n\}, s_0\rangle$, where $s_0, \ldots, s_n \in \mathcal{T}$. Under this construction, the $s_1, \ldots, s_n$ represent the skeletons, and s_0 the erasure of an annotated term. More formally

$$skel(\langle\{s_1, \ldots, s_n\}, s_0\rangle) = \{s_1, \ldots, s_n\} \tag{4.2}$$

$$erase(\langle\{s_1, \ldots, s_n\}, s_0\rangle) = s_0. \tag{4.3}$$

Now, in this setting, we will define rippling as a relation $\Rightarrow_w \subseteq \mathcal{A} \times \mathcal{A}$ where $s \Rightarrow_w t$ iff the following conditions hold:

(i) $erase(s) \Rightarrow erase(t)$,
(ii) $skel(t) \subseteq skel(s)$, and
(iii) $good(s)$ and $good(t)$, where an annotated term is *good* when its skeletons embed into its erasure, i.e.

$$good(\langle\{s_1, \ldots, s_n\}, s\rangle) \equiv \forall i \in [n]. s_i \, \mathcal{E} \, s.$$

By definition, $\Rightarrow_w$ has both the simulation property (the first conjunct) and the skeleton-preservation property (the second conjunct). As defined though, rippling may not terminate. To guarantee this we must add another conjunct (namely that $s > t$, where $>$ is an appropriate well-founded ordering), for example, $s > t$ when the outward bounded wave-fronts in t are "higher" in the skeleton than those in s, as illustrated in Figure 1.3. We will delay formalizing such an ordering and proving that rippling (under this additional requirement) terminates, until Section 4.6.5.

As an example, consider the following sequence of annotated terms where $s \Rightarrow_w t$ and $t \Rightarrow_w r$:

$$s \equiv \langle \{a + b = 42\}, ((c + d) + a) + b = (c + d) + 42 \rangle \qquad (4.4)$$

$$t \equiv \langle \{a + b = 42\}, (c + d) + (a + b) = (c + d) + 42 \rangle \qquad (4.5)$$

$$r \equiv \langle \{a + b = 42\}, (c + d = c + d) \wedge (a + b = 42) \rangle. \qquad (4.6)$$

Equation (4.4), for example, corresponds to the annotated equation

$$(\boxed{(c + d) + a}^{\uparrow}) + b = \boxed{(c + d) + 42}^{\uparrow}. \qquad (4.7)$$

4.3.3 Implementation

It is easy to implement rippling based on the above definition of $\mathcal{A}$ and $\Rightarrow_w$. Annotated terms are represented as pairs where the first component (finite sets) can be implemented, for example, by using lists.

Now suppose we wish to prove some goal g by rippling towards givens $g_1, \dots, g_n$. In the above formalism, we would do this by "lifting" g to an annotated term, s_g, whose skeletons namely are among the g_i, $s_g \equiv \langle \{g'_1, \dots, g'_m\}, g \rangle$ where $\{g'_1, \dots, g'_m\}$ is the largest subset of $\{g_1, \dots, g_n\}$ for which $good(s_g)$ holds. This is trivially computed by testing to see which g_i can be embedded in g. For example, if our given is $a + b = 42$ and goal is $((c + d) + a) + b = (c + d) + 42$, then s is the annotated term given by (4.4).

Now, suppose we have an $s \in \mathcal{A}$ such that $good(s)$ holds. Moreover, suppose there is some term $u \in \mathcal{T}$ such that $erase(s) \Rightarrow u$. Let

$$t = \langle \{s_i \mid s_i \in skel(s) \wedge s_i \, \mathcal{E} \, u\}, u \rangle.$$

If the first component is empty, we reject t, otherwise we have found a t where $s \Rightarrow_w t$.

Based on the observations from Section 4.3.1, it follows that, given a u, we can compute a t (when it exists) in polynomial time. Moreover, for a given s, we can generate all t where $s \Rightarrow_w t$, provided there are only finitely many u where $s \Rightarrow u$ (which is the case whenever $\Rightarrow$ is the rewrite relation defined by a finite set of rewrite rules). Such an implementation of rippling is based on *generate-and-test*, since we first generate candidate terms t and then test whether they satisfy the skeleton-preservation property. As previously noted and illustrated later, we can also enforce termination by testing that $s > t$ under an appropriate ordering $>$.

4.4 Term annotation

4.4.1 The role of annotation

The implementation of rippling just sketched, although potentially effective, is rather inefficient. As in Section 1.4, we consider every possible rewrite step and then analyze the result to see if the givens are preserved and if progress has been made. This is inefficient, since we first generate a candidate rewriting result, and afterwards test the embedding of the givens in the result. If we add annotations to terms, it is possible to combine generation and testing. Term rewriting requires matching, and we can integrate the test for skeleton preservation (and progress) into matching by using annotations to mark what must be preserved, what may change, and how things may change.

To investigate this further, we need to address the question of how to represent annotations on terms. This question is open-ended, since the possibilities for hanging semantic information on terms are also open-ended: how we represent annotations depends on what the annotations should represent and how we intend to use them. This topic, and some of the complexities involved, are considered in detail in Chapter 6. For now, let us concentrate on using annotations to differentiate wave-fronts from skeletons, that is to mark what parts of terms should remain invariant during rewriting.

As observed in Section 1.3, a wave-front can be seen as a "context", i.e. a term with one, or more, proper subterms deleted. Schematically, an outward-directed wave-front is of the form $\boxed{\xi(\,\mu_1,\,\ldots,\,\mu_n\,)}^{\uparrow}$, where $n > 0$ and the μ_i may be similarly annotated. (Inward direct wave-fronts are identical except that the arrow points downwards.) The use of shaded boxes is a useful technique for displaying and visualizing contexts, but such a two-dimensional representation is poorly suited for symbolic manipulation. However, it is simple to develop representations that are easier to formalize and manipulate; below we list several possibilities.

Context markers: Wave-fronts can be represented using markers that state where contexts begin and end.

Symbol markers: Each occurrence of a function symbol in a wave-front can be individually marked, i.e. with an annotation that denotes its status.

Embeddings: The formula is not directly annotated, but the embedding of given into goal is separately recorded.

Under the first approach we might represent the annotated term

$$(\boxed{(c+d)+a}^{\uparrow}) + b = \boxed{(c+d)+42}^{\uparrow}$$

as

$$wf_{up}((c+d) + wh(a)) + b = wf_{up}((c+d) + wh(42)), \qquad (4.8)$$

where wf and wh mark the start and end of contexts, and the subscript up records the direction that the wave-front should move. Wave-fronts with multiple holes correspond to contexts with multiple holes, e.g. multiple occurrences of wh within a wf.

Under the second approach, we might represent the above term by

$$((c^g +^g d^g) +^g a^w) +^w b^w = (c^g +^g d^g) +^g 42^w, \qquad (4.9)$$

where the g superscript means "colored grey" (i.e., in a wave-front), and w means "colored white", i.e., belonging to the skeletons. Different colors can encode different kinds of information. To implement this approach, superscripts can be associated with extra bits, stored with each function symbol occurrence.

Other representations are certainly possible, and there are pros and cons to the various approaches. In this chapter, we will take the first approach and consider an implementation based on it. In Chapter 6, we will look at the second approach.

4.4.2 Formal definitions

Let a signature Σ and set of variables $\mathcal{X}$ be given, and let $\mathcal{T}$ be the set of unannotated terms based on them. We extend Σ with a binary function wf, a unary function wh, and two constant symbols up and dn. The function wf marks the start of a context (the wave-front) and the second argument, either up or dn indicates the *direction* of the wave-front. The function wh marks the end of a context (the wave-hole). Since wf and the direction define related semantic information, we will write this argument of wf as a subscript, i.e. writing $wf_{up}(t)$ as shorthand for $wf(t, up)$. We could also add additional kinds of constructors and annotations (e.g., sinks), but we avoid these details for now.

It is tempting now to define the set of annotated terms $\mathcal{A}$ as freely generated from those functions in the extended signature. However, as noted in Section 4.2.1, we cannot do this, for it would allow too many terms. The constants up and dn alone are meaningless, and not all combinations of wf and wh denote meaningful contexts. Hence, we define $\mathcal{A}$ as a *subset* of the terms in the extended signature, reflecting the role of annotations as context markers.

Definition 1 *Let a signature Σ and set of variables $\mathcal{X}$ be given. The set of annotated terms $\mathcal{A}$ is the smallest set such that*

(i) $t \in \mathcal{A}$ *if* $t \in \mathcal{T}$.

(ii) *For $f \in \Sigma^{(n)}$, $wf_d(f(t_1, \ldots, t_n)) \in \mathcal{A}$ if*
    (a) $d \in \{up, dn\}$,
    (b) $\exists i \in [n]. \exists s \in \mathcal{A}. t_i = wh(s)$; *and*
    (c) $\forall i \in [n]. t_i \in \mathcal{T} \vee \exists s \in \mathcal{A}. t_i = wh(s)$.
(iii) $f(t_1, \ldots, t_n) \in \mathcal{A}$ *if each $t_i \in \mathcal{A}$, $f \in \Sigma^{(n)}$, and $f \notin \{wf, wh, up, dn\}$.*

We use the notation $\mathcal{A}_\Sigma$ to denote those ground annotated terms, $\mathcal{A}_\Sigma \subseteq \mathcal{A}$, *that do not contain variables.*

The interesting case of the definition is the second. Case (ii)(b) states that each wave-front contains at least one immediate subterm that is a wave-hole (i.e., the leading function symbol is wh). Case (ii)(c) states that every immediate subterm must either be unannotated (and therefore cannot contain a wave-hole) or is a wave-hole containing a possibly annotated term, i.e., subterms in wave-fronts are unannotated whereas subterms in wave-holes may be annotated. In Section 4.6.1, we give an example, (4.11), that violates case (ii)(c). Note that, as discussed in Section 2.1.2, insisting that the wave-holes occur as immediate subterms of the function symbol in the wave-front eases the implementation of rippling, since we no longer require wave-front splitting. However, we can reduce clutter (without risking confusion) by displaying wave-fronts in a maximally merged form.

Several additional comments are in order. First, under this definition, a term like $wf_{up}((c + d) + wh(a)) + b$ is an annotated term, but not a term like $wh(a)$, since wave-fronts and wave-holes must be properly nested. This example shows that annotated terms are not closed under the subterm relation. That is, if $s \in \mathcal{A}$, it is not necessarily the case that $s/p \in \mathcal{A}$ for $p \in Pos(s)$. Hence, we see one of the drawbacks of explicitly marking contexts: although annotated terms are terms built over a signature, they are not terms in the conventional sense and one must be careful in applying standard definitions and concepts (e.g. subterm, subterm replacement, etc.) to them.

Second, to simplify the presentation of annotated terms, we will continue using grey shading and arrows in their display. However, this is *just* syntactic sugar for terms built using these additional symbols. Note, moreover, that when the direction associated with a wave-front is unimportant (e.g., in proofs about rippling where the direction is immaterial), we will omit the arrow associated with this notation, e.g., writing $\boxed{s(y)}$ for $\boxed{s(y)}^{\uparrow}$.

Third, much of the complexity in our formalism and subsequent proofs comes from allowing multiple wave-holes, which is necessary for multiple skeletons. We call a wave-front that contains only a single wave-hole *simply annotated*, and it is *multi-hole annotated* otherwise. A term is said to be simply annotated when all its wave-fronts are simply annotated, and is multi-hole annotated otherwise.

Finally, to simplify notation in proofs we will often make a simplifying assumption: we will assume that given a wave-front like $wf_{up}(f(t_1, \ldots, t_n))$, then the arguments of f may be partitioned so that the first $j > 0$ arguments are headed by a wave-hole (i.e., their leading function symbols are wh) and the remaining $n - j$ are not. Hence, the term may be written as $\boxed{f(t_1, \ldots, t_j, t_{j+1}, \ldots, t_n)}^{\uparrow}$, or even $\boxed{f(t_1, \ldots, t_n)}^{\uparrow}$. This is without loss of generality as the proofs we give do not depend on the order of wave-holes.

The skeleton function is defined by recursion on the structure of annotated terms.

Definition 2 *The* skeleton function *skel* $: \mathcal{A} \to \mathcal{P}_1(\mathcal{T})$ *is defined by:*

(i) $skel(X) = \{X\}$, *for all* $X \in \mathcal{X}$.
(ii) $skel(wf_d(f(t_1, \ldots, t_n))) = \{s \mid \exists i \in [n]. t_i = wh(t_i') \wedge s \in skel(t_i')\}$, *for* $d \in \{up, dn\}$.
(iii) $skel(f(t_1, \ldots, t_n)) = \{f(s_1, \ldots, s_n) \mid \forall i \in [n]. s_i \in skel(t_i)\}$, *if* $f \neq wf$.

By erasing annotations, we construct the corresponding unannotated term.

Definition 3 *The* erasure function erase $: \mathcal{A} \to \mathcal{T}$ *is defined by:*

(i) $erase(t) = t$, *for all* $t \in \mathcal{T}$;
(ii) $erase(wf_d(t)) = erase(t)$, *for* $d \in \{up, dn\}$;
(iii) $erase(wh(t)) = erase(t)$;
(iv) $erase(f(t_1, \ldots, t_n)) = f(erase(t_1), \ldots, erase(t_n))$, *if* $f \neq wf$.

Note that the skeleton of a simply annotated term is a singleton set; in this case, we refer to the member as the *skeleton* of the term.

We will apply the skeleton and erasure functions to sets of annotated terms by pointwise application to each element. The erasure function can also be extended to substitutions by

$$erase(\sigma) = \sigma' \text{ where } \sigma'(x) = erase(\sigma(x)).$$

Finally, given a $s \in \mathcal{A}$, we say a subterm t at position p is *in a wave-front in* s if there is a position q above p where the leading function symbol of s/q is wf and there is no position r, $q < r < p$, where the leading symbol of s/r is wh.

As an example, the skeleton of

$$wf_{up}((c + d) + wh(a)) + b = wf_{up}((c + d) + wh(42))$$

is $\{a + b = 42\}$ and the erasure is $((c + d) + a) + b = (c + d) + 42$. The subterms $(c + d) + wh(a)$ and $c + d$ are examples of two terms occurring in wave-fronts.

4.5 Structure-preserving rules

As noted previously, rewrite relations are typically defined by sets of rewrite rules, $s \Rightarrow t \in \mathcal{R}$, where $\mathcal{R} \subseteq \mathcal{T} \times \mathcal{T}$. Suppose such a set $\mathcal{R}$ is given. We now show how to create a set of wave-rules based on $\mathcal{R}$. We continue to ignore the question of termination, and simply focus on the structure preservation requirements.

Definition 4 *A structure preserving rule is given by a pair of terms l and r in $\mathcal{A} \times \mathcal{A}$, written as $l \Rightarrow r$, where*

(i) $erase(l) \Rightarrow erase(r) \in \mathcal{R}$, and
(ii) $skel(r) \subseteq skel(l)$.

Given a set of rewrite rules $\mathcal{R}$, let $\mathcal{W}$ denote the set of all structure-preserving rules based on $\mathcal{R}$. Note that the first requirement tells us that since $erase(l) \Rightarrow erase(r) \in \mathcal{R}$, that $Vars(r) \subseteq Vars(l)$.

The set $\mathcal{W}$ is easily computed. For each $s \Rightarrow t \in \mathcal{R}$ we must find annotated terms $l, r \in \mathcal{A}$ satisfying the above two requirements. Requirement (i) means that l and r correspond to s and t with additional annotations (i.e., wave-fronts and wave-holes). As explained in Section 4.3.1, there are exponentially many possible ways to annotate s and t. Let A_s and A_t be the sets containing these annotations. Then, to satisfy requirement (ii) we form the set of structure-preserving rules $\{l \Rightarrow r \mid l \in A_s \land r \in A_t \land skel(r) \subseteq skel(l)\}$. The set $\mathcal{W}$ is the union of these sets of rules, one such set for each rewrite rule in $\mathcal{R}$.

We call the above procedure for constructing structure-preserving rules a *wave-rule parser*. It is not particularly efficient but, as we will later show in Section 4.8, it can be improved by using a "lazy" procedure where the wave-rules are created on-demand during rewriting itself.

4.6 Rippling using wave-rules

In this section we shall finish our formalization of rippling by explaining how wave-rules are applied, and showing that the resulting rippling relation has the desired properties.

The reader may wonder whether it is necessary to reinvent the wheel here. Our present formalism represents annotated terms as a subset of terms built from an extended signature. Perhaps we can treat these as conventional terms and manipulate these using conventional rewriting? Unfortunately this is not possible. As we have already observed, annotated terms have a special structure, and one cannot directly carry over conventional definitions and results from rewriting.

We consider below the problems involved as they motivate our formalization.

4.6.1 Why ordinary rewriting is not enough

As an example of the problems that can arise, consider applying the wave-rule associated with the recursive definition of times,

$$\boxed{s(U)}^\uparrow \times V \Rightarrow \boxed{(U \times V) + V}^\uparrow, \tag{4.10}$$

to the annotated term $\boxed{s(x)}^\uparrow \times \boxed{s(y)}^\uparrow$. Using conventional rewriting, we match the left-hand side of (4.10) with this term, and this generates the substitutions $\{x/U\}$ and $\{\boxed{s(y)}^\uparrow/V\}$. Performing these replacements in the right-hand side yields

$$\boxed{x \times \boxed{s(y)}^\uparrow + \boxed{s(y)}^\uparrow}^\uparrow .$$

Here we have used a darkened wave-front to indicate an improper nesting; the result is perhaps easier to understand if we desugar the syntax, i.e.

$$wf_{up}(wh(x \times wf_{up}(s(wh(y)))) + wf_{up}(s(wh(y)))). \tag{4.11}$$

The problem is that the (second occurrence of the) subterm $wf_{up}(s(wh(y)))$ is improperly nested within a wave-front: there are two nested wave-fronts without an intermediate wave-hole. Hence, the result of conventional rewriting is a term with annotations, but it is not an annotated term! That is, the result does not belong to the set of annotated terms $\mathcal{A}$.

This problem is simple to explain. Substitution has picked up an annotated term for the variable V. However, annotated terms are not closed under substitution of annotated terms for variables. In particular, we cannot substitute an annotated term for a variable that occurs within a wave-front. This problem, lack of closure under substitution, can be fixed by defining a new notion of substitution, which, in such cases, substitutes the erasure, e.g., yielding

$$x \times \boxed{s(y)}^{\uparrow} + s(y)^{\uparrow}$$

in the above example. However, as this example illustrates, substitutions arise in rewriting from matching, so a redefinition of substitution will require a re-definition of matching. By means of such modifications, we develop a for-malization of rippling that has the desired properties. For unannotated terms and rewrite rules, our formalization of rippling specializes to conventional rewriting.

4.6.2 Ground rippling

We first consider rippling using ground rules. As is typical (see, for example, Dershowitz (1987)), we distinguish between two kinds of variables:

(i) Variables (in rules and, in some cases, in the goal (see Chapter 3)) that can be replaced by substitution.
(ii) Variables (in the goal) that cannot be instantiated by substitution.

We often call variables of the first kind "meta-variables", and write them using upper-case letters, and call variables of the second kind "term variables", which we treat as constants and write using lower-case variables, cf. the discussion of skolemization and dual skolemization in Section 1.2.

Recall that in Section 4.1 we defined $s[t]_p$ to denote the term obtained from s by replacing the subterm at position p by t. Here we extend this to anno-tated terms. For $s, t \in \mathcal{A}$, $p \in Pos(s)$, and $s/p \in \mathcal{A}$ we define annotated replacement, denoted by $s[\![t]\!]_p$ by

$$s[\![t]\!]_p = \begin{cases} s[erase(t)]_p & \text{if } p \text{ is in a wave-front in } s, \\ s[t]_p & \text{otherwise.} \end{cases}$$

For example, replacing a in $\boxed{b + s(a)}^{\uparrow}$ by $\boxed{s(a)}^{\uparrow}$ gives $\boxed{b + s(s(a))}^{\uparrow}$, but

replacing b by $\boxed{s(b)}^{\uparrow}$ gives $\boxed{s(b)}^{\uparrow} + s(a)$. As noted in Section 4.1, sub-stitution can also be seen as a special case of (iterated) term replacement.

Observe that by requiring $s/p \in \mathcal{A}$, we avoid cases where s is an annotated term like $wf_{up}((c + d) + wh(a)) + b$ and p is the position of the subterm $wh(a)$, which is not an annotated term. In such a case, replacement of any

$t \in \mathcal{A}$ at p (i.e., $s[\![t]\!]_p$) would not result in an annotated term. It is easy to check that under the above restrictions, replacement of an annotated subterm by an annotated term always results in an annotated term. Moreover, since unannotated terms are a special case of annotated terms (i.e., $\mathcal{T} \subseteq \mathcal{A}$) and annotated and unannotated replacement agree on this subset, we can without confusion write $s[t]_p$ for $s[\![t]\!]_p$.

In the remainder of this chapter we shall perform all term replacement, including substitutions, using annotated replacement.

Rippling, in the ground case, consists of rewriting using ground structure-preserving rules, i.e., rules that do not contain meta-variables. Let $\mathcal{R}$ be a set of ground rewrite rules and $\mathcal{W}$ a set of ground structure-preserving rules with respect to $\mathcal{R}$. Then we define the corresponding (ground) rippling relation as in conventional rewriting: if $s, l \in \mathcal{A}$, then rippling $s[l]$ at subterm l with the rule $l \Rightarrow r \in \mathcal{W}$ yields $s[r]$. We write $\Rightarrow^G_{\mathcal{W}}$ to denote this (ground) relation and write $s \Rightarrow^G_{\mathcal{W}} t$ to denote that s is transformed to t by *ground rippling* using structure-preserving rules in $\mathcal{W}$.

Ground rippling defines a binary relation on $\mathcal{A} \times \mathcal{A}$ that satisfies the first two properties required of rippling in Section 4.2.2.

Theorem 1 *If $s, t \in \mathcal{A}$ and $s \Rightarrow^G_{\mathcal{W}} t$, then*

(i) $erase(s) \Rightarrow erase(t)$, and
(ii) $skel(t) \subseteq skel(s)$.

Proof (sketch) If $s \Rightarrow^G_{\mathcal{W}} t$, then s, is of the form $s[l]$, t is of the form $s[r]$, and there is a rule $l \Rightarrow r \in \mathcal{W}$. The proof of the above properties follows by structural induction on s. The only non-trivial case occurs when the leading symbol of s is wf, e.g., $s = \boxed{f(s_1, \ldots, s_j, s_{j+1}, \ldots, s_n)}$, and l is a *strict* subterm of one of the s_i. There are two cases, depending on whether $i \leq j$. Consider first the case $i \leq j$. By the induction hypothesis, $erase(s_i[l]) \Rightarrow erase(s_i[r])$. As all the other subterms are unchanged by the replacement, their erasures remain the same. Thus, $erase(s[l]) \Rightarrow erase(s[r])$. From the induction hypothesis, we also have $skel(s_i[r]) \subseteq skel(s_i[l])$. Again, since no other subterm is changed, the union of their skeletons is unchanged and $skel(s[r]) \subseteq skel(s[l])$. If $i > j$, then s_i is an unannotated term within the wave-front. Thus, when we substitute r for l, we will erase annotations on r. Since s_i is unannotated and $l \Rightarrow r$ is a structure-preserving rule, $erase(s_i[l]) \Rightarrow^G_{\mathcal{W}} erase(s_i[r])$ and thus $erase(s[l]) \Rightarrow erase(s[r])$. Finally, from the definition of $skel$ it follows that term replacement in wave-fronts has no effect on the skeletons, so $skel(s[l]) = skel(s) = skel(s[r])$. $\square$

As a simple example, let $\boxed{h(a)}^{\uparrow} \Rightarrow a$ be a structure-preserving rule. We can apply this rule only to the first subterm of the annotated term

$$f(\;\boxed{\boxed{h(a)}^{\uparrow}}\;, h(a))^{\uparrow} \qquad (4.12)$$

because only the first subterm has matching annotations. This application results in $\boxed{f(a, h(a))}^{\uparrow}$. Alternatively, we can apply the rule $a \Rightarrow \boxed{h(a)}^{\uparrow}$ to both occurrences of a in (4.12) resulting in

$$f(\;\boxed{h(\;\boxed{\boxed{h(a)}^{\uparrow}}\;)^{\uparrow}}\;, h(h(a)))^{\uparrow}\;. \qquad (4.13)$$

As required, annotations are erased when substituting $\boxed{h(a)}^{\uparrow}$ for the second occurrence of a. Note too that we can apply this rule arbitrarily often: although structure-preserving rules have the first two properties required of rippling, this example shows that their application is not necessarily terminating; for termination we need further restrictions, which will be introduced later.

4.6.3 Annotated matching

In general, we consider rules that contain (non-term) variables (i.e., $\mathcal{W} \subseteq \mathcal{A} \times \mathcal{A}$), and these are applied using matching.

Definition 5 *For $s, t \in \mathcal{A}$, we call a substitution $\sigma : \mathcal{X} \to \mathcal{A}$ an* annotated match *of s with t iff $Dom(\sigma) = Vars(s)$ and $\sigma(s) = t$.*

As explained previously, we have redefined term replacement and this includes the replacement of variables during substitution. This has important implications for matching. Consider, for example, matching $\boxed{X \times 0}^{\uparrow}$ with $\boxed{s(0) \times 0}^{\uparrow}$. Conventional matching between a pattern s (with variables) and a target t returns, when successful, a unique substitution σ (provided that $Dom(\sigma) = Vars(s)$). However, for this example, both $\{s(0)/X\}$ and $\{\boxed{s(0)}^{\uparrow}/X\}$ are annotated matches.

To make annotated matching feasible, we further restrict it so that there is only a single possible substitution. This restriction is based on defining an ordering on substitutions: if σ_1 and σ_2 are substitutions, we write $\sigma_1 \prec \sigma_2$ iff σ_1 and σ_2 differ on only one variable X, and $\sigma_1(X) = erase(\sigma_2(X))$. We write $\prec^+$ for the transitive closure of $\prec$. Intuitively, this states that $\sigma_1 \prec^+ \sigma_2$ agree,

DELETE:
$$S \cup \{t = t : Pos\} \;\Rightarrow\; S$$

DECOMPOSE:
$$S \cup \{f(s_1, \ldots, s_n) = f(t_1, \ldots, t_n) : Pos\} \;\Rightarrow\; S \cup \{s_i = t_i : Pos \mid 1 \le i \le n\}$$

$$S \cup \{ \boxed{f(s_1, \ldots, s_j, s_{j+1}, \ldots, s_n)} = \boxed{f(t_1, \ldots, t_j, t_{j+1}, \ldots, t_n)} : sk \} \Rightarrow$$
$$S \cup \{s_i = t_i : sk \mid 1 \le i \le j\} \cup \{s_i = t_i : wf \mid j < i \le n\}$$

Figure 4.1 Transformation rules for $amatch(s, t)$.

except that some of the terms $\sigma_1(X)$ are stripped of their annotations. With this in hand, we define when a match is minimal.

Definition 6 *For $s, t \in \mathcal{A}$, then σ is a* minimal match *of s with t iff σ is an annotated match of s with t and there does not exist any annotated match τ with $\tau \prec^{+} \sigma$.*

It follows from this definition that if we have a minimal match, then we cannot remove any annotations and have the result remain a match. It is not difficult to see that minimal matches are unique: substitutions can only differ on variables that occur only in wave-fronts (but not in skeletons), and a minimal match maps these variables to unannotated terms. Hence, we can refer to *the* minimal match of s and t, since it is uniquely defined, provided that there is a match.

We now give an algorithm, $amatch(s, t)$, for computing the minimal match of the pattern s with the target t. It is based on the transformation rules given in Figure 4.1. Because term replacement, and hence substitution, is dependent on context (i.e., whether or not the term to be replaced is in a wave-front), our rules manipulate equations labeled with context information (wf for "in the wave-front" and sk for "in the skeleton"). As notational shorthand, Pos is a meta-variable that matches either $WfName$ or sk.

Starting with the set containing the match problem $\{s = t : sk\}$, we apply these transformation rules exhaustively. An equation set is *reduced* when no transformation rule applies. A reduced equation set S is *compatible* iff

(i) every equation is a variable assignment $X = s : Pos$, for $X \in \mathcal{X}$ and $s \in \mathcal{A}$,

(ii) for each $X \in \mathcal{X}$ there exists at most one equation of the form $X = s : sk$, and

(iii) if $X = s : sk \in S$ and $X = t : wf \in S$ then $erase(s) = t$.

If a reduced equation set is not compatible, matching has failed; otherwise matching has succeeded and we return the answer substitution σ where

$$\sigma(X) = s \text{ iff } X = s:sk \in S \text{ or } (X = s:wf \in S \text{ and } X = t:sk \notin S).$$

As an example, if we match $\boxed{X} + X^{\uparrow}$ with $\boxed{s(a)}^{\uparrow} + s(a))^{\uparrow}$, then our initial matching problem is

$$\{ \boxed{X} + X)^{\uparrow} = \boxed{s(a)}^{\uparrow} + s(a))^{\uparrow} : sk \}$$

and applying DECOMPOSE yields $\{X = \boxed{s(a)}^{\uparrow} : sk, X = s(a) : wf\}$. This reduced equation set is compatible and yields the answer $\{ \boxed{s(a)}^{\uparrow}/X \}$. Note that regular matching would fail on this example.

Some comments are in order. First, the application of these rules in any order terminates in time linear in the size of the smaller of s and t. Second, the DELETE and the first DECOMPOSE rule implement regular matching. For unannotated terms, the compatibility check reduces to the requirement that the reduced equation set contains only variable assignments and each variable has a single substitution. Finally, as with regular matching, we can also add two failure rules for greater efficiency: CONFLICT, which causes annotated matching to fail when the outermost function symbols disagree; and INCOMPATIBLE, which causes annotated matching to fail immediately if the set of equations of the form $X = s:Pos$ is not compatible. These additional failure rules are not, however, needed for the correctness of annotated matching.

The following theorem states that annotated matching functions correctly.

Theorem 2 *For $s, t \in \mathcal{A}$, $amatch(s, t) = \sigma$ iff σ is the minimal match of s and t. When no such match exists, then amatch fails.*

Proof (sketch) The theorem follows from the proof of the following stronger result. Any set of labeled equations S can be transformed to a compatible set of equations with the corresponding answer substitution σ iff for all $s = t:sk \in S$, $\sigma(s) = t$, and for all $s = t:wf \in S$, $erase(\sigma)(s) = t$. The minimality of the answer substitution extracted holds by construction.

$(\rightarrow)$ We perform induction on the length of the transformation. If S is already a reduced compatible set of equations, then the result follows directly from the way the answer substitution is computed. Alternatively, we must apply a transformation rule. The interesting case is when DECOMPOSE is applied, say to $\boxed{f(s_1, \ldots, s_n)}^{\uparrow} = \boxed{f(t_1, \ldots, t_n)}^{\uparrow} : sk$, giving the set of

equations S'. By the induction hypothesis, for all $s = t : sk \in S', \sigma(s) = t$, and
for all $s = t : wf \in S'$, $erase(\sigma)(s) = t$, e.g., $\sigma(s_1) = t_1, \ldots, erase(\sigma)(s_n) = t_n$. It follows that $\sigma(\boxed{f(s_1, \ldots, s_n)}^{\uparrow}) = \boxed{f(t_1, \ldots, t_n)}^{\uparrow}$.

($\leftarrow$) We perform induction over an ordering defined by the multi-set of
the heights of the left-hand sides of the equations. If all left-hand sides are
atomic then, possibly after applications of DELETE, the equation set will be
compatible. If at least one left-hand side is not atomic, then we can pick one
of the form $f(s_1, \ldots, s_n) = t : Pos$ or $\boxed{f(s_1, \ldots, s_n)}^{\uparrow} = t : sk$. In either case,
we can apply DECOMPOSE. The resulting equations are smaller under our
ordering and, hence, we can appeal to the induction hypothesis.

Note that the reduced equation set is not compatible iff either two occurrences of a variable in the skeleton need a different substitution, or a variable
in the wave-front needs a substitution that is not the erasure of the substitution
needed by an occurrence in the skeleton, or there is a conflict in function symbols or annotations preventing application of DECOMPOSE. But this occurs
iff s and t do not have an annotated match. $\qquad\square$

4.6.4 (Non-ground) rippling

We now consider the general case of rippling, which is defined analogously to
conventional rewriting.

Definition 7 *Let $\mathcal{W} \subseteq \mathcal{A} \times \mathcal{A}$ be a set of structure-preserving rules. We say
that s ripples to t, written $s \Rightarrow_w t$, iff there is some $l \Rightarrow r \in \mathcal{W}$, s is of the form
$s[s']$ with $s' \in \mathcal{A}$, there is a substitution $\sigma = amatch(l, s')$, and $t = s[\sigma(r)]$.*

As with normal rewriting, we will write $s[\sigma(l)] \Rightarrow_w s[\sigma(r)]$ to indicate
that $s[\sigma(l)]$ ripples to $s[\sigma(r)]$ under the above conditions.

It is easy to "lift" the results for non-ground rippling to ground-rippling.

Theorem 3 *If $s, t \in \mathcal{A}$ and $s \Rightarrow_w t$, then*

(i) $erase(s) \Rightarrow erase(t)$, and
(ii) $skel(t) \subseteq skel(s)$.

Proof (sketch) Suppose $s, t \in \mathcal{A}$ and there is a rule $l \Rightarrow r \in \mathcal{W}$ for which
$s[\sigma(l)] \Rightarrow_w s[\sigma(r)]$. An easy induction argument on the structure of annotated
terms shows that the erasure and skeleton-preservation properties of structure-
preserving rules are closed under substitution, i.e., $erase(\sigma(l)) \Rightarrow erase(\sigma(r))$,
and $skel(\sigma(r)) \subseteq skel(\sigma(l))$. Now, since $\sigma(l)$ is syntactically identical to a
subterm of s, it is ground (no non-term variables occur in s). Furthermore,

$\sigma(r)$ is also ground because $l \Rightarrow r$ is a structure-preserving rule and thus $Vars(r) \subseteq Vars(l)$. Hence rippling $s[\sigma(l)] \Rightarrow_w s[\sigma(r)]$ is equivalent to rippling $s[\sigma(l)]$ with the ground structure-preserving rule $\sigma(l) \Rightarrow \sigma(r)$. Thus, by Theorem 1, (i) and (ii) hold. □

Let $\Rightarrow_w^*$ be the reflexive transitive closure of $\Rightarrow_w$, i.e., $s \Rightarrow_w^* t$ if s ripples in zero or more steps to t. By induction on the number of steps of rippling, it follows from Theorem 3 that rippling using $\Rightarrow_w^*$ also has these two properties, i.e., if we erase annotations, we can perform the same (object-level) rewriting steps and the annotations merely guide rewriting in a skeleton preserving way.

4.6.5 Termination

We now have a formalization of rippling that has the first two desired properties. Here we address the third property: termination. We define a sufficient condition, based on orderings, for the application of structure-preserving rules to terminate and define wave-rules as those rules that satisfy this condition. In the subsequent section, we give a concrete example of an ordering that has proven useful in practice.

In conventional rewriting, a standard way to show that a term rewriting system given by rules $\mathcal{R}$ terminates is to show that there is a reduction order $>$ that satisfies $l > r$ for all $l \Rightarrow r \in \mathcal{R}$. In the case of annotated terms, an ordering $>$ is a *reduction ordering* when it is:

compatible with contexts: for all $s, l, r \in \mathcal{A}$, if $l > r$, then $s[l] > s[r]$.
closed under substitutions: for all $l, r \in \mathcal{A}$ and $\sigma : \mathcal{X} \to \mathcal{A}$, if $l > r$ then $\sigma(l) > \sigma(r)$.
well-founded: $>$ is well-founded.

The reader should bear in mind here that although this is the standard definition of a reduction ordering,[1] replacement and substitution is that of annotated terms.

We are now finally in a position to define what constitutes a *wave-rule*.

Definition 8 *Let $>$ be a reduction order on annotated terms. Then a structure-preserving rule $l \Rightarrow r$ is a* wave-rule *with respect to $>$ iff $l > r$.*

The condition that $1 > r$ is what we have previously been calling "measure decreasingness", e.g., in Section 1.5.

---

[1] Compatibility with contexts is sometimes replaced in the literature with *closure under Σ-operations*. These are identical (see, for example, Baader and Nipkow (1998)).

With this in hand, we can carry over a standard theorem of term-rewriting into our setting and show that the final property required of rippling holds.

Theorem 4 *For a reduction order $>$ on annotated terms and $\mathcal{W}$ a set of wave-rules with respect to $>$, rippling using wave-rules in $\mathcal{W}$ is terminating.*

Proof Since $>$ is compatible with contexts and closed under substitutions, then for any $t, l, r \in \mathcal{A}$, for which l is a subterm of t, and for any substitution $\sigma : \mathcal{X} \to \mathcal{A}$, we have it that $l > r$ implies $t[\sigma(l)] > t[\sigma(r)]$. Thus, $l > r$ for all $l \Rightarrow r \in \mathcal{W}$ implies $s_1 > s_2$ for all terms s_1, s_2 with $s_1 \Rightarrow_w s_2$. Since $>$ is well-founded, there cannot be an infinite reduction sequence $s_1 \Rightarrow_w s_2 \Rightarrow_w s_3 \cdots$. $\qquad\square$

4.7 Orders on annotated terms

Theorem 4 was proven for an arbitrary reduction order $>$ on annotated terms. In this section, we give a concrete example of a reduction ordering that is useful in practice. We build our ordering in steps, starting with simply annotated terms: those whose wave-fronts have a single wave-hole. We define a measure on such terms and a corresponding order. Afterwards, we generalize this to measures and orders for terms with general (multi-hole) annotations.

4.7.1 Simple annotation

As we have seen in Figure 1.3 in Section 1.4, we can view annotated terms as decorated trees where the tree is the skeleton and the wave-fronts are boxes decorating the nodes. See, for example, the first tree in Figure 4.2, which represents $\boxed{s(U)}^{\uparrow} \geq \boxed{s(V)}^{\uparrow}$. Our orders are based on assigning measures to annotations in these trees. We define orders by progressively simplifying annotated trees to capture the notion of progress during rippling that we wish to measure.

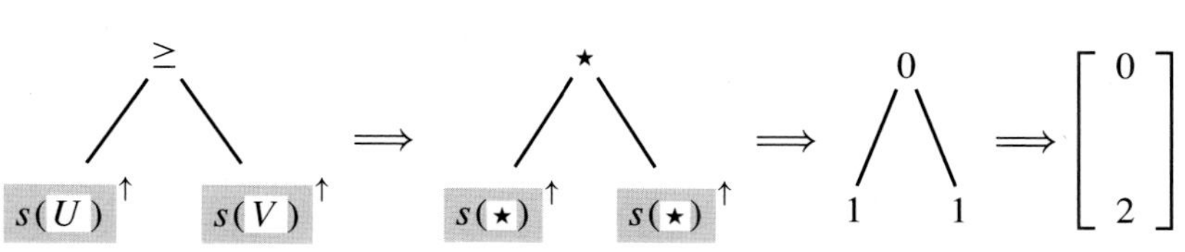

Figure 4.2 Defining a measure on annotated terms.

To begin with, since rippling is skeleton preserving, we need not account for the contents of the skeleton in our orderings. That is, we can abstract away function symbols in the skeleton, for example, mapping each to a variadic function constant "$\star$". This gives, for example, the second tree in Figure 4.2.

A further abstraction is to ignore the names of function symbols within wave-fronts and instead assign a numeric weight to each wave-front. For example, we may tally up the values associated with each function symbol as in a Knuth–Bendix ordering (Knuth & Bendix, 1970). Two of the simplest kinds of weights that we can assign to wave-fronts measure their *width* and their *size*. Width is the number of nested function symbols between the root of the wave-front and the wave-hole. Size is the number of function symbols and constants in a wave-front. In what follows we will restrict our attention to the width measure. This gives, for example, the third tree in Figure 4.2. Of course, there are problem domains where we want our measure to reflect more of the structure of wave-fronts; there we might want to consider other measures, e.g., Bundy (2002a).

Finally, a very simple notion of progress during rippling is that wave-fronts move up or down through the skeleton. Under this view, the tree structure may be ignored: it is not important which branch a wave-front is on, only its depth in the skeleton. Hence, we can apply an abstraction that maps the skeleton tree onto a list, level by level. For instance, we can use the sum of the weights at a given depth. Applying this abstraction gives the final list in Figure 4.2, read bottom to top. *Note that depths are relative to the skeleton as opposed to depth in the erasure.* Measuring depth relative to a fixed skeleton is one of the key ideas in the measure proposed here.

Recall from Section 4.1 that a term has an associated set of positions. If s is a subterm of t at position p, its *depth* is the length of the string p. The *height* of t, written $|t|$, is the maximal depth of any subterm in t. During the remainder of this chapter, positions, depth, and height will *always be relative to the skeleton* of simply annotated terms because we are interested in measures based on weight relative to the skeleton. That is, we picture such terms as in the first tree in Figure 4.2. The positions in the term tree are only those in the skeleton; annotation and function symbols in wave-fronts are treated as markings of function symbols in the skeleton. For example, the term in Figure 4.2 is $\boxed{s(U)}^{\uparrow} \geq \boxed{s(V)}^{\uparrow}$, which has the skeleton $\{U \geq V\}$. The height of this term is 1 since the deepest subterms, U and V, have positions "1" and "2", respectively. Another example is $\boxed{f(s(\boxed{f(a, s(b))}), c)}^{\uparrow}$ with the skeleton $\{f(a, s(b))\}$. The deepest subterm is b at position "21" and, hence, the height of the annotated term is 2.

For an annotated term t, the *out-weight of a position* p is the sum of the weights of the (possibly nested) outwards-oriented wave-fronts at p. We define below a measure on terms corresponding to the final list in Figure 4.2, based on weights of annotations relative to their depths. We also define an analogous measure for inward-directed wave-fronts, except that we reverse the order in which we record their weights in the list, for reasons that will be discussed shortly.

Definition 9 *The* out-measure, $\mathcal{M}^{\uparrow}(t)$, *of an annotated term* t *is a list of length* $|t| + 1$, *the* $(|t| - i)$-*th element of which is the sum of out-weights for all term positions in* t *at depth* i. *The* in-measure, $\mathcal{M}^{\downarrow}(t)$, *is a list, the* $(i + 1)$-*st element of which is the sum of in-weights for all term positions in* t *at depth* i. *The* measure *of a simply annotated term,* $\mathcal{M}(t)$, *is the pair of these measures,* $\langle \mathcal{M}^{\uparrow}(t), \mathcal{M}^{\downarrow}(t) \rangle$.

Consider, for example, the palindrome function ("::" is infix cons)

$$palin(\boxed{H :: T}^{\uparrow}, Acc) \Rightarrow H :: palin(T, \boxed{\boxed{H :: Acc}^{\downarrow}})^{\uparrow} . \tag{4.14}$$

The skeleton of both sides is $\{palin(T, Acc)\}$. The out-measure of the left-hand side is $[1,0]$, and that of the right-hand side is $[0,1]$. The in-measures are $[0,0]$ and $[0,1]$.

We now define a well-founded ordering on these measures that reflects the intended progress of rippling. Consider a simple wave-rule like

$$\boxed{s(U)}^{\uparrow} \times V \Rightarrow \boxed{(U \times V) + V}^{\uparrow} .$$

The left-hand side out-measure is $[1, 0]$, and the right-hand side is $[0, 1]$. Rippling with this rule makes progress because it moves one wave-front upwards towards the root of the term. In general, rippling progresses if one outwards-oriented wave-front moves out or disappears, while nothing deeper moves inwards. If the out-measure of a term before rippling is $[l_0, \ldots, l_k]$ and after $[r_0, \ldots, r_k]$ then there must be some depth d where $l_d > r_d$ and for all i, $0 \leq i < d$ we have $l_i = r_i$. This is simply the lexicographic ordering where components are compared using $>$ on the natural numbers. Progress for inwards-oriented wave-fronts is similar, and reflects that these wave-fronts should move inwards towards leaves. Of course, both outward- and inward-oriented wave-fronts may occur in the same rule, e.g., (4.14). To accommodate both, we define a composite ordering on the out- and in-measures. We order the out-measure before the in-measure, since this enables us to ripple wave-fronts out and either to reach the root of the term, or at some point to turn wave-fronts down and to ripple-in towards the leaves.

Definition 10 $t \succ s$ *iff* $\mathcal{M}(t) > \mathcal{M}(s)$ *and* $skel(s) = skel(t)$. *Here* $>$ *represents the lexicographic extension of* $>_{lex}$ *(the lexicographic order on lists of natural numbers) to pairs.*

This definition is sensible because the restriction that $skel(s) = skel(t)$ means that the measure lists are the same length and may be compared. Although a skeleton-independent measure would be desirable, there is a deeper reason for this restriction: our order would not be closed under substitution without it. As a simple example, consider the terms $s = \boxed{X + s(s(Y))}^{\uparrow}$ and $t = \boxed{s(\boxed{X}) + Y}^{\uparrow}$. If we ignore the skeleton restriction and just compare annotation measures, then $s \succ t$. However, under the substitution $\sigma = \{\boxed{s(s(\boxed{a}))}^{\uparrow}/X\}$ we have $\sigma(t) \succ \sigma(s)$. We shall see that our ordering does not suffer from problems like this.

Given the well-foundedness of $>$ on the natural numbers and that lexicographic combinations of well-founded orders are well-founded, we can conclude:

Theorem 5 *The composite ordering* $\succ$ *is well-founded.*

4.7.2 Multi-hole annotation

We now generalize our order to multi-hole annotation, that is, terms where multiple wave-holes occur in a single wave-front. We have already seen examples of wave-rules involving such terms in Section 2.4, e.g.,

$$binom(\boxed{s(\boxed{X})}^{\uparrow}, \boxed{s(\boxed{Y})}^{\uparrow}) \Rightarrow \boxed{binom(X, \boxed{s(\boxed{Y})}^{\uparrow}) + binom(X, Y)}^{\uparrow}.$$

$$(4.15)$$

Both sides have the same skeleton, namely $\{binom(X, Y)\}$. In general, however, the skeletons of the right-hand side of a wave-rule need only be a subset of the skeletons of the left-hand side.

We define orders for terms with multi-hole annotation in a uniform way from the ordering $\mathcal{M}$ for simply annotated terms by reducing terms with multi-hole annotations to sets of simply annotated terms and extending $\mathcal{M}$ to these sets. This reduction is accomplished by considering ways that multi-hole annotations can be *weakened* to simple annotations by "erasing" wave-holes. We have introduced the concept of weakening in Section 2.4.3; weakening terms with multi-hole annotations means that in a wave-front with multiple wave-holes, some (but not all) of the arguments of the form $wh(t_i)$ are replaced with $erase(t_i)$. A *maximally weak* wave-front is one which has exactly one wave-hole. A *maximally weak* term is one having all its wave-fronts maximally weak. Maximally weak terms are simply annotated and we can apply the previously defined measure $\mathcal{M}$ to them.

Returning to the binomial example, we see that (4.15) has precisely two weakenings:

$$binom(\boxed{s(X)}^{\uparrow}, \boxed{s(Y)}^{\uparrow}) = \boxed{binom(X, \boxed{s(Y)}^{\uparrow}) + binom(X, Y)}^{\uparrow} \quad (4.16)$$

$$binom(\boxed{s(X)}^{\uparrow}, \boxed{s(Y)}^{\uparrow}) = \boxed{binom(X, s(Y)) + binom(X, Y)}^{\uparrow}. \quad (4.17)$$

Both are maximally weak as each wave-front has a single hole. As another example, the left-hand side of (2.23) has four maximal weakenings (and four non-maximal weakenings) while the right-hand side has two weakenings, both maximal.

Let *weakenings*(s) be the set of maximal weakenings of s. It is easily computed by constructing the closure of all weakenings of s, and returning the set of simply annotated results. As elements of these sets are simply annotated, we can apply the measure $\mathcal{M}$ to them. A natural order to define on such sets is therefore the multi-set extension of the order used to compare simply annotated terms. A multi-set extension of an ordering is defined as follows (Dershowitz, 1987).

Definition 11 *Let S be a set and $>$ an ordering, $> \subseteq S \times S$. The multi-set ordering $\gg \subseteq \mathcal{F}(S) \times \mathcal{F}(S)$ induced from $>$ is defined as $M \gg N$ iff N can be obtained from M by replacing one or more elements in M by any finite number of elements each of which is smaller (under $>$) than one of the replaced elements.*

We extend the ordering on simply annotated terms to multi-hole annotated terms as follows.

Definition 12 $l \succ^{\star} r$ *iff* weakenings(l) $\gg$ weakenings(r) *where $\gg$ is the multi-set extension of the order $\succ$.*

This order is well-defined, as maximal weakenings are simply annotated and can be compared using $\succ$. Note that if l and r are simply annotated, then their weakenings are $\{l\}$ and $\{r\}$, and $l \succ^{\star} r$ and $l \succ r$ are equivalent. Hence, we will drop the superscript on $\succ^{\star}$ when context makes our intention clear.

As an example consider (4.15). The left-hand side weakenings are

$$\{binom(\boxed{s(X)}^{\uparrow}, \boxed{s(Y)}^{\uparrow})\}.$$

The right-hand side weakenings are

$$\{\boxed{binom(X, \boxed{s(Y)}^{\uparrow}) + binom(X, Y)}, \boxed{binom(X + s(Y)) + binom(X, Y)}^{\uparrow}\}.$$

The only member of the first set is $\succ$-greater than both members of the second set. This wave-rule is thus measure decreasing.

4.7.3 Termination under $\succ^{\star}$

Since $\succ^{\star}$ is defined via a multi-set extension of a well-founded order, it, too, is well-founded. Hence we immediately have:

Lemma 1 $\succ^{\star}$ *is well-founded.*

We now show that $\succ^{\star}$ is a reduction order. To simplify proofs, we ignore complications caused by inwards-oriented wave-fronts. Reincorporating these is conceptually simple but notationally involved, since measures expand to pairs.

As measures are lists, term replacement corresponds to operations on lists. Hence, we begin with relevant terminology. Let l and r be lists of natural numbers and $l + r$ be componentwise addition. When one list is shorter than the other, we "pad" it out by appending additional zeros to the end, so that its length is the same as the longer list. For n, a natural number, let $l \uparrow^{n}$ be the result of "right shifting" l by n positions by appending l to the end of the list consisting of n zeros. For any natural number d, we define the *splice* of r into l at depth d, which we write as $l +_d r$, to be $l + (r \uparrow^{d})$. Splicing can result in a longer list; for example, if $l = [l_0, l_1, l_2, l_3]$ and $r = [r_0, r_1, r_2]$, then

$$l +_2 r = l + (r \uparrow^2) = [l_0, l_1, l_2, l_3] + [0, 0, r_0, r_1, r_2] = [l_0, l_1, l_2 + r_0, l_3 + r_1, r_2].$$

We will use the following simple properties about splice and list arithmetic below.

Lemma 2 *Let* $l, l', r, r_1, \ldots, r_k$ *be lists of natural numbers and let* $l >_{lex} l'$.

(i) $\forall d \in \mathbb{N}. (l +_d r >_{lex} l' +_d r) \wedge (r +_d l >_{lex} r +_d l')$.

(ii) $\forall d_1 \in \mathbb{N}. \ldots \forall d_m \in \mathbb{N}. (\ldots ((l +_{d_1} r_1) +_{d_2} r_2) \ldots +_{d_m} r_l) >_{lex} (\ldots ((l' +_{d_1} r_1) +_{d_2} r_2) \ldots +_{d_m} r_l)$.

The first part says splicing is monotonic with respect to $>_{lex}$ in both argument positions. The second part is essentially an iterated version of the first for performing multiple splices with different lists at multiple positions. We use these results to prove theorems about closure under substitutions and compatibility with contexts, since such theorems can be seen as statements about splicing measures.

Lemma 3 $\succ^*$ *is compatible with contexts.*

Proof (sketch) Let $s, l, r \in \mathcal{A}$, with l a distinguished subterm of s, i.e., $s[l]$. We must prove that $s[l] \succ^* s[r]$ under the assumption that $l \succ r$. From the assumption, l must be annotated and occurs in the skeleton of s. We argue by cases on the nature of the annotations, and begin with the case where s, l, and r are simply annotated.

In this case, let c be an arbitrary (unannotated) constant, $m_{s[c]} = \mathcal{M}^\uparrow(s[c])$, $m_l = \mathcal{M}^\uparrow(l)$ and $m_r = \mathcal{M}^\uparrow(r)$. Let d be the depth of c in the skeleton of s. Since c is unannotated, the measure of $s[l]$ is the measure of $s[c]$ altered by splicing in at depth d the measure of l, i.e., $m_{s[c]} +_d m_l$. Similarly, the measure of $s[r]$ is $m_{s[c]} +_d m_r$. Since $l \succ r$ we can conclude, using the first part of Lemma 2, that $s[l] \succ^* s[r]$.

Now suppose l and r contain multi-hole annotations and the only multi-hole annotations in $s[l]$ occur in l itself. Let the maximal weakenings of l and r be the sets $L = \{l_1, \ldots, l_j\}$ and $R = \{r_1, \ldots, r_k\}$, respectively. The maximal weakenings of $s[l]$ and $s[r]$ then are the sets $S_l = \{s[l_1], \ldots, s[l_j]\}$ and $S_r = \{s[r_1], \ldots, s[r_k]\}$. Now, under the definition of $\succ^*$ and multi-sets, $l \succ^* r$ if we can replace some collection of the $l_i \in L$ by smaller elements (under $\succ$) resulting in the set R. But we can do the identical replacements in the context $s[\cdot]$ hence transforming the set S_l to S_r. Consider such a replacement, say replacing $l_1 \in L$ by $r_1, \ldots, r_p$; now $l_1 \succ r_i$ and it follows (by the previously considered case) that $s[l_1] \succ s[r_i]$ for each $i \in \{1, \ldots, p\}$. Hence the transformation of S_l to S_r shows that $s[l] \succ^* s[r]$.

The final case to consider is when s itself has multiple skeletons, independent of the number of skeletons of l. We argue as above, except that rather than just comparing sets composed from $s[l_i]$ and $s[r_i]$ we have to consider weakenings of s as well. But any steps in weakening s (not in the subterm l) can be made identically in both $s[l_i]$ and $s[r_i]$ and $s[l] \succ^* s[r]$ follows. $\qquad\square$

Lemma 4 $\succ^*$ *is closed under substitutions.*

Proof (sketch) Let s and t be in $\mathcal{A}$ with $s \succ^* t$. To show that $\sigma(s) \succ^* \sigma(t)$ we can, without loss of generality, consider a substitution σ that replaces a single variable x with some $r \in \mathcal{A}$. We consider two cases: first, when s, t are simply annotated, and second, when they may contain multi-hole annotations.

Case 1: Terms s and t are simply annotated. As $s \succ t$, both terms have the same single skeleton. Note that substitutions for occurrences of x in wave-fronts have no effect on our width measure (although they can change the size of a wave-front). Assume x occurs p times in each skeleton. If $weakenings(r) = \{r_1, \ldots, r_m\}$ then

$$S = weakenings(\sigma(s)) = \{s_1, \ldots, s_n\}$$

and

$$T = weakenings(\sigma(t)) = \{t_1, \ldots, t_n\},$$

where $n = p * m$. Each of these weakenings can be constructed by replacing the variables x in s and t with maximal weakenings of r; each s_i thus has a "partner", t_i, in which the occurrences of x are replaced by the same weakening of r. Now to show that S is greater than T under the multi-set ordering we must give a transformation of S to T where each term is replaced by a finite number of smaller (under $\succ$) terms. Our transformation is simply to replace s_i by its partner, t_i. If we order (arbitrarily) the occurrences of x in the skeleton of s (and therefore also t), $x_1, \ldots, x_p$, then if s_i and t_i were formed by replacing x_j, occurring at depth d_j with a weakening of t that has a measure r_j, then the measures of the two terms s_i and t_i are

$$(\ldots((s +_{d_1} r_1) +_{d_2} r_2) \ldots +_{d_p} r_p)$$

and

$$(\ldots((t +_{d_1} r_1) +_{d_2} r_2) \ldots +_{d_p} r_p),$$

respectively. But now, using the second part of Lemma 2, we have that the former is greater under $>_{lex}$ than the latter; hence, $\sigma(l) \succ \sigma(r)$.

Case 2: All terms may contain multi-hole annotations. Let $S = \{s_1, \ldots, s_j\}$ and $T = \{t_1, \ldots, t_k\}$ be the maximal weakenings of s and t. As $s \succ^* t$, there is a transformation (respecting $\succ$) of S to T. We must construct a transformation from the maximal weakenings of $\sigma(s)$ to the maximal weakenings of $\sigma(t)$. We proceed as follows. Consider a replacement of, say, s_1 in S with some $t_1, \ldots, t_p$ that takes place in transforming S to T. Now suppose the maximal weakenings of r are $\{r_1, \ldots, r_m\}$. Then $\sigma(s_1)$ and the $\sigma(t_i)$ each have n maximal weakenings where n is a multiple of m dependent on the number of occurrences of x in the skeleton of s_1. In particular, $weakenings(\sigma(s_1)) = \{s_{1,1}, \ldots, s_{1,n}\}$ and for each t_i, $weakenings(\sigma(t_i)) = \{t_{i,1}, \ldots, t_{i,n}\}$. Again we may speak of "partners": each $s_{1,j}$ has as partners $t_{i,j}$, for $i \in \{1, \ldots, p\}$ and $j \in \{1, \ldots, n\}$ where the weakenings of $t_{i,j}$ come from weakening the occurrences t identically to their weakenings in $s_{1,j}$. Furthermore, because for each $i \in \{1, \ldots, p\}$, $s_1 \succ t_i$, we can use case 1 to conclude that each maximal weakening of $\sigma(s_1)$ is larger than its partners. Hence replacing each $s_{1,i}$ with its partners defines an appropriate transformation from *weakenings*$(\sigma(s))$ to *weakenings*$(\sigma(t))$. □

From the previous lemmas we can conclude that $\succ^\star$ is a reduction ordering on annotated terms and, hence, by Theorem 4 we have:

Theorem 6 *Rippling using wave-rules $l \Rightarrow r$, with respect to $\succ^\star$, is terminating.*

4.8 Implementing rippling

We have completed our formal account of rippling and termination orderings for annotated terms. We now turn to the more practical problem of mechanizing rippling. In particular, given an ordering, how do we recognize wave-rules and apply them? Below we describe an implementation of the above formalization of rippling that is part of the Edinburgh *C L^AM* system. To give the reader a feel for this, and the issues involved, we sketch briefly a couple of the core routines.

Much of the work in implementing rippling concerns turning unannotated rewrite rules into wave-rules; as noted in Section 4.5 we call this *wave-rule parsing*. A wave-rule parser takes a set of unannotated rewrite rules and returns a corresponding set of wave-rules. By definition, the wave-rules are annotated copies of the original rules that are skeleton preserving and measure decreasing. We can achieve these requirements separately. An *annotation phase* first annotates l and r with unoriented wave-fronts (i.e., making no commitment to whether the direction is *up* or *dn*) in a skeleton-preserving way. The annotation algorithm works by calculating the maximal skeletons that can be embedded in both sides of the wave-rule, and encloses the remaining expressions in wave-fronts. Afterwards, an *orientation phase* augments each wave-front with a direction so that $l \succ^\star r$. The result is a wave-rule or, more accurately, a set of wave-rules, since there are choice-points in both phases.

As an example, consider parsing a rewrite rule such as

$$s(U) \times V \Rightarrow (U \times V) + V. \tag{4.18}$$

We may proceed by annotating this so the two sides have identical skeletons, e.g.

$$\boxed{s(U)} \times V \Rightarrow \boxed{(U \times V) + V} \, . \qquad (4.19)$$

Afterwards we can orient the annotations yielding the wave-rule

$$\boxed{s(U)}^{\uparrow} \times V \Rightarrow \boxed{(U \times V) + V}^{\uparrow}. \qquad (4.20)$$

Both sides of (4.20) now have the same skeleton, and the measure of the left-hand side is greater than that of the right-hand side.

Any implementation, however, must cope with the problem that under our definition of wave-rules, a given rewrite rule can generate exponentially many (in the size of the input rule) wave-rules. Computing and storing all possible wave-rules is expensive both in time and space and complicates efficient wave-rule lookup. For example, in the previous example, there are other possible legal parsings such as:

$$\boxed{s(U) \times V}^{\uparrow} \Rightarrow \boxed{U} \times V + V^{\downarrow} \qquad (4.21)$$

$$s(U) \times \boxed{V}^{\uparrow} \Rightarrow U \times \boxed{V} + V^{\downarrow} \qquad (4.22)$$

$$\boxed{s(U) \times V}^{\uparrow} \Rightarrow U \times V + \boxed{V}^{\downarrow}. \qquad (4.23)$$

These additional parsings are problematic, as there are often many of them. However, they are admissible under our definition, and even find use on rare occasion, e.g. in unblocking. Rather than trying to say in advance which wave-rules could be useful in practice, and thereby should be returned by the parser, our solution to this problem is to compute wave-rules dynamically, by parsing "on demand". We describe this in the following section.

4.8.1 Dynamic wave-rule parsing

The implementation of rippling in *CLAM* is based on a *dynamic wave-rule parser* that, given a data-base of unannotated rewrite rule, applies them by annotating them only as required during rewriting. That is, given a term $s[t]$ to be rippled, we look for an unannotated rule $l \Rightarrow r$ where l matches the erasure of t. When this is the case, $l \Rightarrow r$ is a candidate rewrite rule that can be further processed to become a wave-rule. We proceed by computing annotations for l that allow for an annotated match with t; afterwards, based on these annotations and their orientations, we compute annotations and orientations for r so that $l \Rightarrow r$ is a wave-rule.

```
ripple(T,NT) :-              % ripple at some term position
  subterm(Pos,T,ST),         % pick subterm ST in T at position Pos
  pick_rule(L,R),            % pick a rule L -> R
  match_rule(L,R,ST,NR),     % can rule be annotated to match ST
  replace(Pos,T,NR,NT).      % replace ST with NR yielding NT

match_rule(L,R,ST,NR) :-
  copy_an(ST,L,AL),          % copy annotations from ST onto L
  amatch(AL,ST,Sigma),       % annotated match of AL with ST
  parse(AL,R,AR),            % find annotations for R
  apply_subs(Sigma,AR,NR)    % apply sub to AR yielding NR

parse(AL,R,AR) :-
  pick_an(R,A),              % annotate R
  skel_preserving(AL,A),     % skeletons equal?
  orient(AL,A,AR).           % orient R
```

Figure 4.3 Wave-rule parser (top-level routines).

Figure 4.3 contains the Prolog program (we assume here that the reader is familiar with Prolog) that implements the top-level routines for rippling based on dynamic parsing. We illustrate the procedure through an example. Suppose we wish to perform one step of rewriting of the term T, given by $\boxed{s(x)}^{\uparrow} \times$ $\boxed{s(y)}^{\uparrow}$. Moreover, suppose that our collection of unannotated rewrite rules includes (4.18), the recursive definition of multiplication, which is defined by

$$s(U) \times V \Rightarrow (U \times V) + V.$$

`ripple` picks a subterm ST of T and a rule $L \Rightarrow R$. In our example, a solution to this (there may be others that are returned on backtracking) is where ST is T itself and the rule selected is the above one. In dynamic parsing we need only generate annotations for the right-hand sides of wave-rules whose left-hand sides arise during the proof. This is performed in `match_rule`, which starts by copying annotations from ST onto L; this yields AL, an annotated version of L. Copying annotations fails if ST and L have erasures that do not match. In our example, $AL = \boxed{s(U)}^{\uparrow} \times V$. The program finds an annotated match of AL with ST, generating a suitable substitution for the rewriting step. In our example, we get the substitution x for U and $\boxed{s(y)}^{\uparrow}$ for V. Afterwards, `parse` is called to find an annotation of R with the same skeleton as AL, and with a maximum[1] orientation, in this case $\boxed{(U \times V) + V}^{\uparrow}$.

---

[1] Maximum under our order. When there are multiple choices with the same measure, the program returns all of them on backtracking.

The substitution is then applied to this annotated right-hand side, yielding
$(x \times \boxed{s(y)}^{\uparrow})^{\uparrow} + s(y)$. Note that substitution is that for annotated terms;
regular substitution would generate an improperly annotated result. The final
step in `ripple` replaces the rewritten subterm within the context of the su-
perterm from which it came, again using subterm replacement for annotated
terms.

The supporting subroutines for parsing are fairly straightforward. We used
generate (`pick_an`) and test (`skel_preserving`) to generate skeleton-
preserving annotations of the right-hand side of rewrite rules. This takes exponential time but the efficiency can be considerably improved by interleaving
generation and testing (i.e., testing during generation) or using dynamic programming. In our experience, naïve generate-and-test has acceptable performance.

The routine `orient` finds an orientation of the wave-fronts on the right-
hand side that yields a measure smaller than the left-hand side. This can be
implemented naïvely by generating orientations (there are two possibilities for
each wave-front) and comparing the two sides of the proposed rule under the
given measure. By comparing possible orientations against each other, we can
return the maximum possible right-hand side orientations. As with annotation,
there are algorithms to implement orientation more efficiently. In practice, it is
often the case that all annotations are simple (single wave-holes), and then it is
possible to orient the right-hand side in linear time.

4.8.2 Sinks and colors

One kind of annotation we have not discussed in our measures or parsing is
sinks. This is deliberate, as we can safely ignore sinks in both the measure
and the parser. Sinks only serve to decrease the applicability of wave-rules by
creating additional preconditions; that is, we only ripple inwards if there is a
sink or nested wave-front within the wave-front (see Section 2.3). Hence sinks
decrease the search space of rippling, and termination without this restriction
implies termination with this restriction. The value of sinks is that they restrict
search without reducing the *utility* of rippling: their use guides rippling in a
way that allows the givens to be successfully used.

In Section 2.4, we also considered using different colors to distinguish
different skeletons. The motivation behind the introduction of colors is that
rippling only preserves a subset of the skeletons, and colors help prevent us
ending up with the wrong subset. Since colored rippling is a restriction of

uncolored rippling, termination follows immediately from termination in the uncolored case. Colors thus also increase the utility of rippling. Although colors are not needed for showing the termination of rippling, they actually implicitly arose in our discussion on termination. Each color has a single skeleton and the reduction order defined in Section 4.7 compares the measures of different colors separately.

Colored skeletons also suggest an alternative formal account of wave annotation. Colors essentially allow us to ripple simultaneously towards several different givens (or towards the same given in more than one way): each distinct ripple corresponding to a different color. Therefore, we could replace multi-holed wave-fronts by multiple single-holed annotations, each of a different color. This would greatly simplify the wave-measure, since only simply annotated terms would be required. However, we would then need to take into account multiple simple wave-measures at each ripple step. We would need to check that none of these wave-measures increased, and that at least one decreased. This would effectively reintroduce the machinery of multi-sets of wave-measures.

5

The scope and limitations of rippling

In this chapter we survey applications of rippling, both within and outwith induction, successful and unsuccessful, by a wide variety of researchers. We start with examples where rippling has successfully guided a proof attempt and reduced the amount of search. We will see that rippling is both applicable and successful in a surprisingly wide variety of situations.

However, we also survey some failures of rippling. The notion of failure is a fuzzy one. In many of the cases discussed, we speculate on extensions of rippling that might turn failure into success. We also give examples of where proof critics can analyze the failure, and suggest a successful patch to the proof attempt.

Section headers of successful examples are preceded by "Hit" and unsuccessful ones by "Miss".

5.1 Hit: bi-directionality in list reversal

The following rather artificial example was constructed to illustrate the ability of rippling to use rewrite rules in both orientations while still guaranteeing termination:

$$\forall K : list(\tau). \forall : list(\tau). qrev(qrev(K, L), [\,]) = rev(L) <> rev(rev(K))).$$

The details of the step case of the inductive proof can be found in Figure 5.1. Note that rewrite rules (5.7) and (5.8) are the same equation oriented in opposite directions. They are applied at steps (5.3) and (5.2), respectively.

This example also illustrates that rippling is not affected by redundant rules. Suppose the wave-rule

$$(\boxed{X <> Y}^{\uparrow}) <> Z \Rightarrow X <> (\boxed{Y <> Z}^{\downarrow}) \tag{5.1}$$

Given:

$$qrev(qrev(tl, L), [\ ]) = rev(L) <> rev(rev(tl))$$

Goal and ripple:

$$qrev(qrev(\boxed{[hd|tl]}^{\uparrow}, \lfloor l \rfloor), [\ ]) = rev(\lfloor l \rfloor) <> rev(rev(\boxed{[hd|tl]}^{\uparrow}))$$

$$qrev(qrev(tl, \lfloor [hd|l] \rfloor), [\ ]) = rev(\lfloor l \rfloor) <> rev(\boxed{rev(tl) <> [hd]}^{\uparrow}) \qquad (5.2)$$

$$qrev(qrev(tl, \lfloor [hd|l] \rfloor), [\ ]) = rev(\lfloor l \rfloor) <> \boxed{rev([hd]) <> rev(rev(tl))}^{\uparrow}$$

$$qrev(qrev(tl, \lfloor [hd|l] \rfloor), [\ ]) = \boxed{rev(\lfloor l \rfloor) <> rev([hd])}^{\downarrow} <> rev(rev(tl))$$

$$(5.3)$$

$$qrev(qrev(tl, \lfloor [hd|l] \rfloor), [\ ]) = rev(\lfloor [hd] <> l \rfloor) <> rev(rev(tl))$$

$$qrev(qrev(tl, \lfloor [hd|l] \rfloor), [\ ]) = rev(\lfloor [hd|l] \rfloor) <> rev(rev(tl))$$

Wave-rules:

$$qrev(\boxed{X :: Y}^{\uparrow}, Z) \Rightarrow qrev(Y, \boxed{X :: Z}^{\downarrow}) \qquad (5.4)$$

$$rev(\boxed{X :: Y}^{\uparrow}) \Rightarrow \boxed{rev(Y) <> X :: nil}^{\uparrow} \qquad (5.5)$$

$$X <> (\boxed{Y <> Z}^{\uparrow}) \Rightarrow (\boxed{X <> Y}^{\uparrow}) <> Z \qquad (5.6)$$

$$\boxed{rev(K) <> rev(L)}^{\downarrow} \Rightarrow rev(\boxed{L <> K}^{\downarrow}) \qquad (5.7)$$

$$rev(\boxed{L <> K}^{\uparrow}) \Rightarrow \boxed{rev(K) <> rev(L)}^{\uparrow} \qquad (5.8)$$

Figure 5.1 Bi-directionality in a list-reversal example. The rippling proof starts with the applications of the recursive definitions of *qrev* and *rev* (wave-rules (5.4) and (5.5)) to the left- and right-hand sides, respectively. The lemmas (5.8), (5.6), and (5.7) are then applied, in turn, to the right-hand side. Wave-rules (5.7) and (5.8) are weakened during application, to merge one of their wave-holes into the wave-front. Finally, the sink on the right-hand side is simplified to make it equal to the one on the left-hand side.

were available. Wave-rule (5.1) is the associativity of $<>$, but oriented in the opposite direction to (5.6) in Figure 5.1. Unannotated, wave-rule (5.1) would apply at step (5.3), undoing the work of wave-rule (5.6). However, the mismatch of the wave annotation prevents this application. It is not possible to annotate rule (5.1) as a wave-rule so that it could undo the work of wave-rule (5.6).

Note that rippling can be used to apply lemmas like (5.6), (5.8), and (5.7) as well as recursive definitions like (5.4) and (5.5).

5.2 Hit: bi-conditional decision procedure

The next example illustrates that rippling scales up to a large problem requiring inductive proof. It arises in the synthesis of a decision procedure for the bi-conditional fragment of propositional logic (Armando, *et al.* 1998). Rippling does not completely automate the proof, but it does succeed in automating the proofs of all the key lemmas. One such lemma requires six inductions and four generalizations, which rippling completes without backtracking. The proof provides a nice illustration of the interplay of rippling, fertilization, and generalization. This lemma is too long to give here, so we have selected a simpler one for our illustration. The lemma

$$\forall s.\forall w.\Delta(\eta(s, w)) \vdash \forall w.\forall w'.\Delta(\forall s.(s \in w \to \eta(s, w')))$$

is proved by structural induction on w, where $\Delta(A)$ means A is decidable and $s \in A$ means s occurs in A. The proof of the step case can be found in Figure 5.2 on p136.

The same group also completed an even harder case study of the synthesis of a unification algorithm, with a similarly successful performance from rippling (Armando, *et al.* 1999).

5.3 Hit: reasoning about imperative programs

Andrew Ireland and Jamie Stark (Ireland & Stark, 1997; 2001; Stark & Ireland, 1998) have used rippling to reason about the correctness of imperative programs within the context of a Floyd–Hoare-style logic (Floyd, 1967; Hoare, 1969). Specifications in this logic are triples, $\{P\}C\{Q\}$, where P and Q denote predicates, while C is imperative program code; P and Q are known as the *precondition* and *postcondition*, respectively. The operational interpretation of $\{P\}C\{Q\}$ is as follows. If C is executed within a state where the program variables satisfy precondition P, then, on the termination of C, the program variables will satisfy postcondition Q. To illustrate, consider *exp* given in Figure 5.3 on p137 for computing exponentiation. The standard approach to verifying imperative code is to propagate pre-conditions and post-conditions so that every atomic statement is sandwiched between two assertions. Once this is achieved, a set of *verification conditions* can be extracted automatically, which collectively implies the correctness of the code with respect to the given assertions. A verification condition is a purely logical statement and hence it can be tackled directly using general-purpose theorem-proving techniques. In the case of the exp example, the verification condition generated for the **while**-loop takes the form:

$$r * exp(x, y) = exp(\mathcal{X}, \mathcal{Y}) \wedge (y > 0) \rightarrow$$
$$(r * x) * exp(x, y - 1) = exp(\mathcal{X}, \mathcal{Y}).$$

The proof of this verification conjecture can be presented in terms of rippling, as shown in Figure 5.4 on p137.

Part of the given associated with such verification conditions (e.g. $r * exp(x, y) = exp(\mathcal{X}, \mathcal{Y})$), is known as the *loop invariant*. This particular form is called a *tail-invariant*. The challenge of imperative program verification comes from the fact that discovering loop invariants is in general undecidable. Heuristics for discovering loop invariants are well documented within the literature (Gries, 1981; Kaldewaij, 1990; Katz & Manna, 1976) and the problem is essentially the same as that of generalization encountered within proof by mathematical induction. For this reason, the technique for generalization described in Section 3.5 can also be used to guide the discovery of tail-invariants, e.g. the one given above. In addition, the basic idea underlying the induction revision proof patch (see Section 3.3) has also been shown to have relevance for the *replacement of constants* heuristic found within the literature (Stark & Ireland, 1998). The novelty of this approach is that the discovery and verification of loop invariants goes hand-in-hand, with the ripple method leading the way.

5.4 Hit: lim+ theorem

Rippling has also been applied outwith inductive proofs. In this example we see how it has been applied to theorems from analysis.

Lim+ is the limit theorem stating that "the limit of the sum is the sum of the limits". It can be formalized as

$$\forall f_1 : \mathbb{R} \mapsto \mathbb{R}. \forall f_2 : \mathbb{R} \mapsto \mathbb{R} \lim_{X \mapsto A} (f_1(X) + f_2(X))$$
$$= \lim_{X \mapsto A} f_1(X) + \lim_{X \mapsto A} f_2(X),$$

where

$$\lim_{X \mapsto A} f(X) = L \leftrightarrow$$
$$\forall \epsilon.(0 < \epsilon \rightarrow \exists \delta.(0 < \delta \wedge \forall X.(X \neq A \wedge |X - A| < \delta \rightarrow |f(X) - L| < \epsilon))).$$

Interest in this theorem was sparked by Bledsoe *et al.* (1972), who introduced a "limit heuristic" to guide the proofs for a number of similar theorems, of which Lim+ was the simplest. Lim+ then became a challenge problem for uniform proof procedures without the use of the "limit heuristic" (Bledsoe, 1990). It proved to be very challenging. The resolution theorem prover Otter (McCune, 1991), for instance, can only find a proof of the simplest of the six axiomatizations in Bledsoe (1990), and even then it requires user guidance to restrict function-nesting. Totally automatic proofs of the simplest two axiomatizations have been found, for instance, by Digricoli's RUE prover; RUE can also find proofs of the remaining four axiomatizations if a small amount of user interaction is allowed (Digricoli, 1994).

A rippling proof for Lim+ is given in Yoshida *et al.* (1994). As this proof is non-clausal it is not directly comparable to any of the axiomatizations in Bledsoe (1990). The idea for using rippling on this theorem is due to Woody Bledsoe. Interestingly, rippling seems to subsume some of the ideas implicit in the "limit heuristic". The rippling proof uses wave-rules loosely based on the clausal axiomatization of Lim+ given in Bledsoe (1990) and requires no search. The definitions of lim must first be unpacked. The limit of the sum is treated as the goal and the limits of the two summands as the givens. A difference unification algorithm (Basin & Walsh, 1993) (which combines the computation of an embedding, as in Section 4.3.1, with a first-order unifier) is used to annotate the goal with wave-fronts. The subsequent rippling proof is shown in Figure 5.5 on p138, and the wave-rules used in this proof are given in Figure 5.6 on p139.

5.5 Hit: summing the binomial series

Rippling has also been used in several different ways to sum series (Walsh *et al.*, 1992). For instance, the *perturbate* method makes an incremental change to the sum, similar to an induction step, and then ripples the incremented sum against itself. The *standard form* method ripples the sum against one or more previously solved sums. We illustrate the standard form method below with a binomial series problem.

The series to be summed is

$$\sum_{i=0}^{n} \binom{i+1}{m+1}, \tag{5.17}$$

where $\binom{n}{r}$ is the number of combinations of r things from n. The closed form of this sum is found with the aid of the previously solved sum

$$\sum_{i=0}^{N} \binom{i}{M} = \binom{N+1}{M+1}. \tag{5.18}$$

Difference-unifying the goal (5.17) against the given, i.e. the left-hand side of (5.18), annotates (5.17) as

$$\sum_{i=0}^{n} \left(\left\lfloor \frac{\boxed{i}+1^{\uparrow}}{\boxed{m}+1^{\uparrow}} \right\rfloor \right).$$

The subsequent rippling proof and the wave-rules used are given in Figure 5.7 on p140. After two fertilizations with the given, this yields the following closed form for the sum

$$\binom{n+1}{m+2} + \binom{n+1}{m+1}.$$

5.6 Hit: meta-logical reasoning

Santiago Negrete (1994; 1996) has used rippling to reason in framework logics. The rules of inference of an object-logic are represented as axioms in a framework or meta-logic, in his case Edinburgh LF (Harper *et al.*, 1992). A theorem to be proved is first split into given and goal using introduction rules of natural deduction calculi. The given and goal are then difference-unified, which inserts wave annotations in both. Rippling is then used to reduce the difference between them. An example from a simple propositional logic is given in Figure 5.8 on p141. Several other examples from other logics can be found in Negrete's thesis (Negrete, 1996). This example illustrates the use of rippling on both the given and the goal.

5.7 Hit: SAM's lemma

Jürgen Cleve and Dieter Hutter have used rippling to prove SAM's lemma (Cleve & Hutter, 1994); SAM's lemma is a theorem known as Bumcroft's identity in lattice theory:

> *Let a and b be two elements of a modular lattice with unique complements, and $\overline{a}$*
> *and $\overline{b}$ their unique complements. Assuming the join and the meet of a and b exist,*
> *does the join of $\overline{a}$ and $\overline{b}$ exist, and is it the unique complement of the meet of a and*
> *b?*

It was the first open conjecture to be proved by an automated theorem prover, called SAM, hence the name, although user interaction was required (Guard *et al.*, 1969). It is used still as a standard test problem for resolution theorem provers.

Let $\sqcap$ and $\sqcup$ be two associative, commutative, and idempotent functions satisfying the absorption rules:

$$X \sqcup (X \sqcap Y) = X \qquad X \sqcap (X \sqcup Y) = X. \qquad (5.23)$$

Furthermore, there are top and bottom elements (0 and 1) of the lattice which is assumed to be modular:

$$X \sqcap Y \rightarrow X \sqcap (Y \sqcup Z) = Y \sqcup (X \sqcap Z). \qquad (5.24)$$

Finally, complements are defined by

$$comp(X, Y) \Leftrightarrow (X \sqcap Y = 0 \wedge X \sqcup Y = 1). \qquad (5.25)$$

Then SAM's lemma is stated as

$$comp(a, c \sqcup d) \wedge comp(b, c \sqcap d) \rightarrow (a \sqcup (b \sqcap c)) \sqcap (a \sqcup (b \sqcap d)) = a. \qquad (5.26)$$

Cleve and Hutter used nested applications of rippling within the INKA theorem prover. The axioms and hypotheses are turned into wave-rules. The goal to be proved is difference unified against a given, which is heuristically selected from the axioms and hypotheses. Rippling is then used to reduce the difference between them. If this ripple is blocked, then the term immediately containing the blocked wave-front is difference unified against one side of a heuristically selected equation. Rippling then proceeds to reduce the difference between them. This process may nest several levels deep before a proof is found. An example of the rippling used on one level is given in Figure 5.9 on p142.

5.8 What counts as a failure?

In the remainder of this chapter we survey some failures of rippling, but first we discuss what we mean by this.

As we have seen above, rippling is applicable wherever a *goal* is to be proved with the aid of a structurally similar *given*, so it would be inappropriate to count as failures the many situations in which there is no structurally similar given to ripple towards. We restrict our attention to situations in which there *are* such givens and the goal *can* be rewritten so that some form of fertilization is possible, but rippling *fails* to produce the required rewriting. We can summarize this by saying that there is a *fertilization proof* but no *rippling proof*.

In many cases, it will clarify the rippling failure to attempt unsuccessfully to annotate the rewrite rules or the goals. Ill-annotations of this kind will be preceded by *, similar to the way that linguists mark ungrammatical sentences. The text will then explain why the annotation fails.

5.9 Miss: mutual recursion

It has been noted for some time that rippling fails for inductive proofs about mutual recursive functions, and a wide variety of solutions have been proposed, e.g. abstraction of the skeleton, temporary relaxation of skeleton preservation, deriving wave-rules from mutually recursive definitions, generalization of the goal, cycling the skeleton preservation between several givens, etc. Many of these solutions involve extensions to rippling.

Here is a simple example of the problem. Consider the following two versions of the *even* predicate, one defined using mutual recursion and the other without it:

$$even_m(0) \Rightarrow true$$
$$odd_m(0) \Rightarrow false$$
$$* \qquad even_m(\boxed{s(N)}^{\uparrow}) \Rightarrow odd_m(N) \qquad\qquad (5.27)$$
$$* \qquad odd_m(\boxed{s(N)}^{\uparrow}) \Rightarrow even_m(N) \qquad\qquad (5.28)$$

$$even_r(0) \Rightarrow true$$
$$even_r(s(0)) \Rightarrow false$$
$$even_r(\boxed{s(s(N))}^{\uparrow}) \Rightarrow even_r(N).$$

Note that the two *ed rules fail as wave-rules due to non-preservation of skeletons.

Suppose we try to prove these two definitions equivalent, namely

$$\forall x{:}nat.\ even_m(x) \leftrightarrow even_r(x).$$

The step case of a 2-step induction provides a non-ripple fertilization proof

$$even_m(\boxed{s(s(x))}^{\uparrow}) \leftrightarrow even_r(\boxed{s(s(x))}^{\uparrow})$$

$$* \qquad odd_m(\boxed{s(x)}^{\uparrow}) \leftrightarrow even_r(x) \qquad\qquad (5.29)$$

$$even_m(x) \leftrightarrow even_r(x).$$

This fails to be a rippling proof because skeleton preservation fails in line (5.29), where the skeleton of the left-hand side is $odd_m(x)$ instead of $even_m(x)$.

This particular example yields to a simple solution: derive some real wave-rules by resolving (5.27) and (5.28) together.

$$even_m(\boxed{s(s(N))}^{\uparrow}) \Rightarrow even_m(N) \qquad\qquad (5.30)$$

$$odd_m(\boxed{s(s(N))}^{\uparrow}) \Rightarrow odd_m(N) \qquad\qquad (5.31)$$

An alternative rippling proof is available. However, this solution does not work in general. Consider, for instance, mutual recursive definitions of the form

$$f_1(\boxed{s(N)}^{\uparrow}) \Rightarrow \boxed{g_1(N,\ f_1(N),\ f_2(N))}^{\uparrow}$$

$$f_2(\boxed{s(N)}^{\uparrow}) \Rightarrow \boxed{g_2(N,\ f_1(N),\ f_2(N))}^{\uparrow}.$$

A more general-purpose solution is to adapt the definition of a skeleton so that mutually recursive functions are abstracted to the same equivalence class. Within the skeleton, $even_m$ and odd_m would be abstracted to the same equivalence class; the abstracted skeleton would now be preserved during the proof, which could then be directed by rippling.

5.10 Miss: commuted skeletons

To deal with mutually recursive functions it is not always necessary to abstract the skeleton to preserve it. The following example shows that a lost skeleton can sometimes be restored.

Consider the theorem

$$\forall n{:}nat.\ even_m(n) \lor odd_m(n),$$

where $even_m$ and odd_m are as defined in Section 5.9. The step case of a proof of this theorem by structural induction gives the non-rippling fertilization proof

$$even_m(\lfloor s(n)\rfloor) \lor odd_m(\lfloor s(n)\rfloor)$$

$$* \qquad odd_m(n) \lor even_m(n) \qquad\qquad (5.32)$$

$$even_m(n) \lor odd_m(n). \qquad\qquad (5.33)$$

This fails as a rippling proof because the skeleton is disrupted in line (5.32). However, commuting it in line (5.33) restores the original skeleton and allows fertilization. The concept of rippling proof might be extended to include this example if skeleton preservation were defined modulo the commutativity of $\lor$. Doubtless there are similar examples involving other commutative functions, predicates, or connectives. Note that the alternate wave-rules, (5.30) and (5.31), will yield an alternative rippling proof in this case, but this solution is not available in general.

5.11 Miss: holeless wave-fronts

Another problem is caused by ill-formed wave-fronts. Consider the following rewrite rules, which cannot be annotated as wave-rules.

$$* \ length(W) = 6 \to split\_list(\; \boxed{H :: T}^{\uparrow}, W\;)$$

$$\Rightarrow \boxed{W :: split\_list(T, \; \boxed{H :: nil}^{\downarrow})}^{\uparrow}$$

$$* \ D = 6 \to new\_split(\; \boxed{H :: T}^{\uparrow}, W, D\;)$$

$$\Rightarrow \boxed{W :: new\_split(T, \; \boxed{H :: nil}^{\downarrow}, \boxed{1}^{\downarrow})}^{\uparrow}$$

The function *split_list* takes a list and splits it into a list of sublists of 6 elements each (or possibly less than 6 for the last one). The first argument is the input list, and the second is an accumulator of the current sublist; *new_split* is a more efficient version of the same function, which keeps a running count of the current sublist length in its third argument, rather than having to recalculate it each time. The two rules above arise from corresponding cases of the definitions of these two functions. In these cases, the current sublist has reached length 6. This sublist is then attached to the output and a new sublist is started.

These rules both fail to be wave-rules because their annotation is ill-formed; there are no wave-holes in their inward wave-fronts. Without such wave-holes, no skeletons can be defined for the right-hand sides. This phenomenon usually happens in the context of accumulators. The inductive proof still goes through because the free variable arising from the accumulator is able to match with the holeless wave-front, as in the example below.

A non-rippling fertilization proof arises in the proof of

$$\forall x{:}list,\ w{:}list.\ new\_split(x, w, length(w)) = split\_list(x, w).$$

In the case $length(w) = 6$ of the step case, we get the proof steps

$$new\_split(\boxed{h :: t}^{\uparrow}, \lfloor w \rfloor, length(\lfloor w \rfloor)) = split\_list(\boxed{h :: t}^{\uparrow}, \lfloor w \rfloor)$$

$$* \ new\_split(\boxed{h :: t}^{\uparrow}, \lfloor w \rfloor, length(\lfloor w \rfloor)) = \boxed{w :: \boxed{split\_list(t, \boxed{h :: nil}^{\downarrow})}}^{\uparrow}$$

$$new\_split(h :: t, w, length(w)) = w :: new\_split(t, h :: nil, length(h :: nil))$$

$$w :: new\_split(t, h :: nil, 1) = w :: new\_split(t, h :: nil, 1).$$

This fails to be a rippling proof: (a) because the rewrite rules used are not wave-rules, and (b) because skeleton preservation cannot be checked since it is not clear what the skeletons are in the presence of sinks.

One possible solution is based on a revised definition of sink annotation and, hence, skeleton. Each sink is labeled by the meta-variable to which it corresponds in the given. This label is used to form the skeleton. This enables us to annotate the two rules above as wave-rules, that is

$$length(W) = 6 \rightarrow split\_list(\boxed{H :: T}^{\uparrow}, \lfloor W \rfloor_W)$$

$$\Rightarrow \boxed{W :: \boxed{split\_list(T, \lfloor H :: nil \rfloor_W)}}^{\uparrow}$$

$$D = 6 \rightarrow new\_split(\boxed{H :: T}^{\uparrow}, \lfloor W \rfloor_W \lfloor D \rfloor_D)$$

$$\Rightarrow \boxed{W :: \boxed{new\_split(T, \lfloor H :: nil \rfloor_W, \lfloor 1 \rfloor_D)}}^{\uparrow}.$$

The skeletons on each side of these two rules are *split_list(T, W)* and *new_split(T, W, D)*, respectively, so the skeleton is preserved in both rules. This change of annotation also turns the above proof steps into the following rippling proof.

$$new\_split(\boxed{h :: t}^{\uparrow}, \lfloor w \rfloor_W, length(\lfloor w \rfloor_W)) = split\_list(\boxed{h :: t}^{\uparrow}, \lfloor w \rfloor_W)$$

$$new\_split(\boxed{h :: t}^{\uparrow}, \lfloor w \rfloor_W, length(\lfloor w \rfloor_W)) = \boxed{w :: \boxed{split\_list(t, \lfloor h :: nil \rfloor_W)}}^{\uparrow}$$

$$new\_split(h :: t, w, length(w)) = w :: new\_split(t, h :: nil, length(h :: nil))$$

$$w :: new\_split(t, h :: nil, 1) = w :: new\_split(t, h :: nil, 1).$$

We are not aware of any disadvantages of this modified definition of sink annotation.

5.12 Miss: inverting a tower

The following non-rippling fertilization proof arises in the application of linear logic to recursive planning (Cresswell *et al.*, 1999).

Linear logic allows reasoning about limited resources. Resources are represented as hypotheses of the conjecture, whose proof is sought. Hypotheses representing limited resources cannot be used in a proof more than once. In planning applications, the conjecture asserts that the goal state can be reached from the initial state, where the reachability is asserted using the linear logic implication $\multimap$. A concrete plan can then be recovered from a proof of the conjecture.

Consider the conjecture

$$\forall t. \forall a.\ tower(t) \otimes hn \otimes tower(a) \tag{5.34}$$
$$\multimap tower(empty) \otimes tower(revput(t, a) \otimes hn),$$

where the following rewrite rules are available

$$*\quad tower(\boxed{H :: T}^{\uparrow}) \otimes hn \Rightarrow tower(T) \otimes hold(H) \tag{5.35}$$

$$*\quad hold(H) \otimes tower(T) \Rightarrow hn \otimes tower(\boxed{H :: T}^{\downarrow}) \tag{5.36}$$

$$revput(\boxed{H :: T}^{\uparrow}, A) \Rightarrow revput(T, \boxed{H :: A}^{\downarrow}). \tag{5.37}$$

These rewrite rules define the operators available to the recursive planner. Here, *tower(l)* means that *l* is a tower of blocks; *hn* means that the robot's hand is free; *revput*(*l*, *a*) is the list *a* with the reverse of the list *l* appended to its front, and *hold*(*h*) means that the robot is holding block *h*. So:

- Rule (5.35) describes the robot's ability to pick up the top block of a tower.
- Rule (5.36) describes the robot's ability to place a block it is holding on top of a tower.
- Rule (5.37) is the step case of a recursive definition of *revput*.

However, note that rule (5.35) is not skeleton preserving and rule (5.36) is neither skeleton preserving nor measure decreasing.

The non-rippling fertilization step case of the proof of (5.34) by list induction is

$$(tower(\boxed{h :: t}^{\uparrow}) \otimes hn) \otimes tower(\lfloor a \rfloor)$$

$$\multimap tower(empty) \otimes (tower(revput(\boxed{h :: t}^{\uparrow}, \lfloor a \rfloor)) \otimes hn$$

∗ $\quad (tower(t) \otimes hold(h)) \otimes tower(\lfloor a \rfloor)$

$$\multimap tower(empty) \otimes (tower(revput(t, \lfloor h :: a \rfloor)) \otimes hn$$

∗ $\quad tower(t) \otimes (hold(h) \otimes tower(\lfloor a \rfloor))$

$$\multimap tower(empty) \otimes (tower(revput(t, \lfloor h :: a \rfloor)) \otimes hn$$

∗ $\quad tower(t) \otimes (hn \otimes tower(\lfloor h :: a \rfloor))$

$$\multimap tower(empty) \otimes (tower(revput(t, \lfloor h :: a \rfloor)) \otimes hn$$

∗ $\quad (tower(t) \otimes hn) \otimes tower(\lfloor h :: a \rfloor)$

$$\multimap tower(empty) \otimes (tower(revput(t, \lfloor h :: a \rfloor)) \otimes hn.$$

Each of the ∗ed steps fails to be skeleton preserving.

A partial solution is available if we move the *hn* and *hold* expressions from the skeletons to the wave-fronts. Rules (5.35) and (5.36) become

$$\boxed{tower(\boxed{H :: T}^{\uparrow}) \otimes hn}^{\uparrow} \Rightarrow \boxed{tower(T) \otimes hold(H)}^{\uparrow} \tag{5.38}$$

∗ $\quad \boxed{hold(H) \otimes tower(T)}^{\downarrow} \Rightarrow \boxed{hn \otimes tower(\boxed{H :: T}^{\downarrow})}^{\uparrow} , \tag{5.39}$

although rule (5.36) still fails to be measure decreasing.

The step case becomes

$$(\boxed{tower(\boxed{h :: t}^\uparrow)} \otimes hn \;)^\uparrow \otimes tower(\lfloor a \rfloor)$$

$$\multimap tower(empty) \otimes (\boxed{tower(revput(\boxed{h :: t}^\uparrow, \lfloor a \rfloor))} \otimes hn \;)^\uparrow$$

$$(\boxed{tower(t) \otimes hold(h)}^\uparrow) \otimes tower(\lfloor a \rfloor)$$

$$\multimap tower(empty) \otimes (\boxed{tower(revput(t, \lfloor h :: a \rfloor))} \otimes hn \;^\uparrow)$$

$$tower(t) \otimes (\boxed{hold(h) \otimes \boxed{tower(\lfloor a \rfloor)}}^\downarrow)$$

$$\multimap tower(empty) \otimes (\boxed{tower(revput(t, \lfloor h :: a \rfloor))} \otimes hn \;^\uparrow)$$

$$* \qquad tower(t) \otimes (\boxed{hn \otimes tower(\lfloor h :: a \rfloor)}^\uparrow)$$

$$\multimap tower(empty) \otimes (\boxed{tower(revput(t, \lfloor h :: a \rfloor))} \otimes hn \;^\uparrow)$$

$$* \qquad (\boxed{tower(t) \otimes hn}^\uparrow) \otimes tower(\lfloor h :: a \rfloor)$$

$$\multimap tower(empty) \otimes (\boxed{tower(revput(t, \lfloor h :: a \rfloor))} \otimes hn \;^\uparrow).$$

The last two steps still fail to be measure decreasing, but the skeleton-preservation problems are fixed.

The above problems are not specific to linear logic, but are similar to the problems with mutual recursion, described in Section 5.9, in that hn and $hold(h)$ are closely related (e.g. hn could be defined as $\forall h. \neg hold(h)$), and the rewrite rules tend to switch between these alternative representations disrupting the skeleton. More of the skeleton would be preserved if hn was replaced by $\forall h. \neg hold(h)$.

5.13 Miss: difference removal

The various difference-removing techniques, such as E-resolution (Morris, 1969), RUE resolution (Digricoli, 1979) and equality graphs (Bläsius & Siekmann, 1988), provide a family of non-rippling fertilization proofs. In difference removal, wave-fronts are transformed into skeleton rather than rippled out of the way. To illustrate this, consider the conjecture

$$* \qquad (\forall w. \forall z.\ p(\ \boxed{g(a, w)}^{\uparrow},\ \boxed{z}^{\uparrow})) \rightarrow p(\ \boxed{f(e)}^{\uparrow},\ \boxed{b}^{\uparrow}),$$

where the following equations are available as rewrite rules:

$$g(X, X) = h(X, b), h(U, V) = h(V, U), h(b, a) = f(b), b = c,\ \text{and}\ c = e.$$

This yields the following non-rippling fertilization proof

$$* \qquad p(\ \boxed{f(e)}^{\uparrow},\ \lfloor b \rfloor)$$

$$* \qquad p(\ \boxed{f(c)}^{\uparrow},\ \lfloor b \rfloor)$$

$$* \qquad p(\ \boxed{f(b)}^{\uparrow},\ \lfloor b \rfloor)$$

$$* \qquad p(\ \boxed{h(b, a)}^{\uparrow},\ \lfloor b \rfloor)$$

$$* \qquad p(\ \boxed{h(a, b)}^{\uparrow},\ \lfloor b \rfloor)$$

$$p(g(a, \lfloor a \rfloor), \lfloor b \rfloor).$$

At this point, the goal can be fertilized with the given.

This fails to be a rippling proof for two reasons: the ill-formedness caused by the holeless wave-fronts, and the non-decrease of the wave-measure. As discussed in Section 5.11, the definition of wave annotation can be readily extended to include holeless wave-fronts. We have also experimented with a wave-measure that can reward the transformation of wave-fronts into skeletons as well as the moving of wave-fronts in desirable directions (Bundy, 2002b). These wave-measures succeed in integrating difference removal with difference moving (rippling), but at the cost of non-termination.

5.14 Best-first rippling

We have experimented with a best-first approach to rippling, which provides a general solution to many of the rippling failures discussed above. By regarding the preservation of the skeleton and a decrease in the wave-measure as being merely heuristics for preferring one rewrite over another, we can tolerate either of these deviations from rippling when no rippling step is available, or when all rippling steps have been tried and failed. In the worst-case, rippling will degrade gracefully into conventional rewriting. Thus, all non-rippling fertilization proofs will be encompassed within the wider remit of best-first rippling.

The down-side of best-first rippling is that the search space is significantly increased, although rippling provides some heuristics to control the increased search. It is also less obvious when to fire the critics, since rippling will rarely completely fail. Critic preconditions have to be recast as heuristics that may now fire in preference to some of the weaker forms of rippling.

5.15 Rippling implementations and practical experience

We conclude this chapter with some remarks on the implementations of rippling and our experiences with the resulting systems.

Rippling has been implemented in two different ways: the Edinburgh $CLAM$ family uses context markers and the Saarbrücken INKA system uses symbol markers. For more than 10 years, these systems have been used in various academic and industrial case-studies. In the early 1990s, INKA was integrated into the VSE system, which is a tool for formal software development. Since its first release in 1994, VSE has been used in various industrial applications including a scheduling system for distributed radio broadcasting, access control software for nuclear power plants, security models for various smartcard applications, and communication and security protocols. In each of these industrial case-studies, typically half of the development time was spent using the VSE tool to prove the proof obligations that arose. As a consequence, thousands of proofs have been completed, many of which were inductive proofs concerned with the properties of specified abstract data types. In more than 95 per cent of these inductive proofs, rippling is successful, provided that appropriate lemmas (to generate appropriate wave-rules) are available. Hutter (1997) gives an overview of the time the system needs to prove various inductive theorems. Five years later, making use of hardware improvements, the system typically needed only between 0.001 and 0.1 s to prove each of the theorems Hutter mentions. Even if rippling fails to enable the use of the induction hypothesis, this failure usually becomes obvious in under 1 s. Thus, rippling can be implemented efficiently enough for large practical applications. Rippling imposes severe constraints on the proof search, and some of these restrictions may be lifted to enlarge the search space in order to find other rippling-related proofs. For example, we may allow the system to manipulate wave-fronts to unblock rippling.

In terms of industrial strength loop invariant verification, rippling has been successfully applied (Ireland *et al.*, 2004) to SHOLIS (King *et al*, 2000). SHOLIS is a safety-critical application which was developed, using the

SPARK approach (Barnes, 2003), to meet the UK Ministry of Defense Interim Defense Standards 00-55 and 00-56.

Some improvements have been implemented in order to overcome combinatorial explosions during the generation of wave-rules. For instance, the specification of a record data type rec results in an equation

$$\forall x : rec. f(g_1(x), \ldots, g_n(x)) = x$$

with $g_1, \ldots, g_n$ denoting the access functions to each of the fields of the record x. Computing the wave-rules in a naïve way results in about $2^n - 1$ different wave-rules depending on which occurrences of x are considered as part of the skeleton. As a consequence, weakening of a skeleton was built into the matching algorithm, so that the wave-rule with the largest possible skeleton $\boxed{f(g_1(x), \ldots, g_n(x))} = x$, replaces all other possible wave-rules.

The failures of rippling in these examples are typically due to:

- missing wave-rules, because the specification of the theory lacked appropriate axioms or lemmas;
- the use of an inappropriate induction ordering or a missing case analysis that would enable the application of a suitable conditional wave-rule; or
- the problem that recursive function definitions (especially using mutual recursion) do not satisfy the syntactical requirements to generate wave-rules.

In many of these cases, critics have been developed to remedy these failures.

5.16 Summary

In this chapter we have illustrated rippling, both successful and unsuccessful applications, on a series of examples. These examples have served to make the following points. First, rippling is widely applicable both within and outwith inductive proof. Second, rippling improves on symbolic evaluation by the application of (recursive) definitions because it is also able to use previously proved lemmas. Third, rippling improves on conventional rewriting to normal form in three respects: (a) some undesirable rewrites are prevented by the wave annotation; (b) rippling is able to use rewrite rules in *both* orientations without loss of termination; and (c) rippling has a uniform termination proof that does not require modification when new wave-rules are added. Finally, when rippling fails, an analysis of the partial proof is often able to suggest a suitable patch, e.g. in the form of speculating a missing lemma or a generalization of the induction formula (cf. Chapter 3).

However, there are still some non-rippling fertilization proofs, that is proofs where we would like rippling to apply but it appears not to. Fortunately, there is hope. Deeper analysis suggests that there are ways to extend rippling in order to encompass these proofs as rippling proofs. Among the techniques that enable us to extend rippling, best-first rippling stands out as having wide applicability. It is also necessary to extend wave annotation to permit holeless wave-fronts. Wave annotation via embeddings is also important as a way of extending rippling to higher-order logic.

Some of the rippling extensions proposed as solutions to rippling failures come at a price, namely increased search. It may be possible to recognize when these extensions are needed and only use them when it is necessary to do so. For instance, mutual recursion can be recognized syntactically, and rippling can be adapted dynamically to cope with it.

Givens:

$$\Delta(\forall s.(s \in w_1 \rightarrow \eta(s, w'))) \qquad \Delta(\forall s.(s \in w_2 \rightarrow \eta(s, w')))$$

Goal and ripple:

$$\Delta(\forall s.(s \in \boxed{w_1 \iff w_2}^{\uparrow} \rightarrow \eta(s, w')))$$

$$\Delta(\forall s.(\boxed{s \in w_1 \vee s \in w_2}^{\uparrow} \rightarrow \eta(s, w')))$$

$$\Delta(\forall s.(\boxed{s \in w_1 \rightarrow \eta(s, w') \wedge s \in w_2 \rightarrow \eta(s, w')}^{\uparrow}))$$

$$\Delta(\boxed{\forall s.(s \in w_1 \rightarrow \eta(s, w')) \wedge \forall s.(s \in w_2 \rightarrow \eta(s, w'))}^{\uparrow})$$

$$\boxed{\Delta(\forall s.(s \in w_1 \rightarrow \eta(s, w'))) \wedge \Delta(\forall s.(s \in w_2 \rightarrow \eta(s, w')))}^{\uparrow}$$

Wave-rules:

$$S \in \boxed{W_1 \iff W_2}^{\uparrow} \Rightarrow \boxed{S \in W_1 \vee S \in W_2}^{\uparrow} \tag{5.9}$$

$$\boxed{A \vee B}^{\uparrow} \rightarrow C \Rightarrow \boxed{A \rightarrow C \wedge B \rightarrow C}^{\uparrow} \tag{5.10}$$

$$\forall x.(\boxed{A \wedge B}^{\uparrow}) \Rightarrow \boxed{\forall x.A \wedge \forall x.B}^{\uparrow} \tag{5.11}$$

$$\Delta(\boxed{A \wedge B}^{\uparrow}) \Rightarrow \boxed{\Delta(A) \wedge \Delta(B)}^{\uparrow} \tag{5.12}$$

Figure 5.2 A rippling sequence from a decision procedure synthesis.

$\iff$ is the constructor of bi-conditional formulas. The proof proceeds by the application of wave-rules (5.9), (5.10), (5.11), and (5.12). Wave-rule (5.9) is part of the definition of $\in$. Wave-rules (5.10), (5.11), and (5.12) are domain-independent logical wave-rules. Sink markers have been omitted to avoid clutter, since sinks are not used in this example.

exp: $\{x = \mathcal{X} \wedge y = \mathcal{Y}\}$
    $r := 1;$
    while $(y > 0)$ **do**
    begin
        $r := r * x;$
        $y := y - 1$
    end
    $\{r = exp(\mathcal{X}, \mathcal{Y})\}$

Figure 5.3 Algorithm for computing exponentiation: x, y, and r denote program variables while $\mathcal{X}$ and $\mathcal{Y}$ are constants, sometimes referred to as *ghost variables*. Ghost variables are used within the post-condition to refer to the initial values of programs variables. The function *exp*, which appears within the post-condition, is defined as follows:

$$Y = 0 \rightarrow exp(X, Y) = 1$$
$$Y > 0 \rightarrow exp(X, Y) = X * exp(X, Y - 1)$$

Given:

$$r * exp(x, y) = exp(\mathcal{X}, \mathcal{Y})$$

Goal and ripple:

$$\boxed{(r * x)}^{\uparrow} * exp(x, \boxed{y - 1}^{\uparrow}) = exp(\mathcal{X}, \mathcal{Y})$$

$$r * \boxed{(x * exp(x, \boxed{y - 1}^{\uparrow}))}^{\downarrow} = exp(\mathcal{X}, \mathcal{Y})$$

$$r * exp(x, y) = exp(\mathcal{X}, \mathcal{Y})$$

Wave-rules:

$$\boxed{(X * Y)}^{\uparrow} * Z \Rightarrow X * \boxed{(Y * Z)}^{\downarrow}$$

$$Y > 0 \rightarrow \boxed{X * exp(X, \boxed{Y - 1}^{\uparrow})}^{\downarrow} \Rightarrow exp(X, Y)$$

Figure 5.4 The invariance of a loop invariant. The given is the value of the loop invariant before a traversal of the while loop and the goal is its value afterwards. The wave-rules are derived from the associativity of multiplication and the definition of *exp*.

Givens:

$$\forall \varepsilon.(0 < \varepsilon \to \exists \delta.(0 < \delta \land \forall x.(x \neq a \land |x - a| < \delta \to |f_1(x) - l_1| < E)))$$

$$\forall E.(0 < E \to \exists \delta.(0 < \delta \land \forall x.(x \neq a \land |x - a| < \delta \to |f_2(x) - l_2| < E)))$$

Goal and ripple:

$$\forall \varepsilon.(0 < \varepsilon \to \exists \delta.(0 < \delta \land \forall x.(x \neq a \land |x - a| < \delta \to |\,\boxed{f_1(x) + f_2(x)} - \boxed{l_1 + l_2}\,| < \lfloor \varepsilon \rfloor)))$$

$$\forall \varepsilon.(0 < \varepsilon \to \exists \delta.(0 < \delta \land \forall x.(x \neq a \land |x - a| < \delta \to \boxed{f_1(x) - l_1 + f_2(x) - l_2} < \lfloor \varepsilon \rfloor)))$$

$$\forall \varepsilon.(0 < \varepsilon \to \exists \delta.(0 < \delta \land \forall x.(x \neq a \land |x - a| < \delta \to \boxed{|f_1(x) - l_1| + |f_2(x) - l_2|} < \lfloor \varepsilon \rfloor)))$$

$$\forall \varepsilon.(0 < \varepsilon \to \exists \delta.(0 < \delta \land \forall x.(x \neq a \land |x - a| < \delta \to \boxed{|f_1(x) - l_1| < \lfloor \tfrac{\varepsilon}{2} \rfloor} \land \boxed{|f_2(x) - l_2| < \lfloor \tfrac{\varepsilon}{2} \rfloor}))$$

$$\forall \varepsilon.(0 < \varepsilon \to \exists \delta.(0 < \delta \land \boxed{\forall x.(x \neq a \land |x - a| < \delta \to |f_1(x) - l_1| < \lfloor \tfrac{\varepsilon}{2} \rfloor) \land \forall x.(x \neq a \land |x - a| < \delta \to |f_2(x) - l_2| < \lfloor \tfrac{\varepsilon}{2} \rfloor)}))$$

$$\forall \varepsilon.(0 < \varepsilon \to \boxed{\exists \delta.(0 < \delta \land \forall x.(x \neq a \land |x - a| < \delta \to |f_1(x) - l_1| < \lfloor \tfrac{\varepsilon}{2} \rfloor)) \land \exists \delta.(0 < \delta \land \forall x.(x \neq a \land |x - a| < \delta \to |f_2(x) - l_2| < \lfloor \tfrac{\varepsilon}{2} \rfloor))})$$

$$\forall \varepsilon.(\boxed{0 < \lfloor \varepsilon \rfloor} \to \boxed{\exists \delta.(0 < \delta \land \forall x.(x \neq a \land |x - a| < \delta \to |f_1(x) - l_1| < \lfloor \tfrac{\varepsilon}{2} \rfloor))} \land \boxed{0 < \lfloor \varepsilon \rfloor} \to \exists \delta.(0 < \delta \land \forall x.(x \neq a \land |x - a| < \delta \to |f_2(x) - l_2| < \lfloor \tfrac{\varepsilon}{2} \rfloor))$$

$$\boxed{\forall \varepsilon.(0 < \lfloor \varepsilon \rfloor \to \exists \delta.(0 < \delta \land \forall x.(x \neq a \land |x - a| < \delta \to |f_1(x) - l_1| < \lfloor \tfrac{\varepsilon}{2} \rfloor))} \land \boxed{\forall \varepsilon.(0 < \lfloor \varepsilon \rfloor \to \exists \delta.(0 < \delta \land \forall x.(x \neq a \land |x - a| < \delta \to |f_2(x) - l_2| < \lfloor \tfrac{\varepsilon}{2} \rfloor)))}$$

Figure 5.5 The lim theorem. Note how the wave-fronts move steadily outwards during the rippling proof exposing more and more of the skeleton in the wave-holes and decreasing the amount of skeleton outside the wave-holes. This example illustrates several aspects of rippling. There are two givens, the goal being rippled towards both of them simultaneously. Most of the wave-fronts are rippled-out to the top level, but some get rippled-sideways into two of the sinks. Note that the final fertilization is prevented because not all sinks have the same content. This mismatch strongly motivates the relevant patch, which is to equalize the contents of the sinks using $0 < X \leftrightarrow 0 < \frac{X}{2}$.

$$\boxed{X + Y}^{\uparrow} - \boxed{A + B}^{\uparrow} \Rightarrow \boxed{(X - A) + (Y - B)}^{\uparrow} \tag{5.13}$$

$$|\,\boxed{X + Y}^{\uparrow}\,| \Rightarrow \boxed{|X| + |Y|}^{\uparrow} \tag{B-8}$$

$$\boxed{X + Y}^{\uparrow} < E \Rightarrow \boxed{X < \boxed{\frac{E}{2}}^{\downarrow} \wedge Y < \boxed{\frac{E}{2}}^{\downarrow}}^{\uparrow} \tag{B-11.3}$$

$$A \rightarrow \boxed{B \wedge C}^{\uparrow} \Rightarrow \boxed{(A \rightarrow B) \wedge (A \rightarrow C)}^{\uparrow} \tag{5.14}$$

$$\forall X.\ \boxed{A \wedge B}^{\uparrow} \Rightarrow \boxed{\forall X.\ A \wedge \forall X.\ B}^{\uparrow} \tag{5.15}$$

$$[\forall X.\forall Y.\ A(X) \wedge 0 < Y < X \rightarrow A(Y)] \wedge$$
$$[\forall X.\forall Y.\ B(X) \wedge 0 < Y < X \rightarrow B(Y)] \rightarrow$$
$$\exists X.0 < X \wedge \boxed{A \wedge B}^{\uparrow} \Rightarrow \boxed{\exists X.0 < X \wedge A \wedge \exists X.0 < X \wedge B}^{\uparrow} \tag{5.16}$$

Figure 5.6 Wave-rules used to prove lim+.

These are the wave-rules used to prove Lim+ in Figure 5.5. They are applied in order except for the last two steps, which apply wave-rules (5.14) and (5.15) again. The numbers (B-n) by some wave-rules correspond to some of the clauses in Bledsoe (1990). Note that there can be no corresponding rules for the logical wave-rules, (5.14) and (5.15) since Bledsoe's axioms are in clausal form. These two wave-rules correspond to part of the process of clausification itself. Wave-rules (5.13) and (5.16) are special to this problem. Note that the value of X on the left-hand side of (5.16) is the minimum of its two values on the right-hand side. This rule is related to clause 10.3 in Bledsoe (1990).

Given: $\sum_{i=0}^{N} \binom{i}{M}$

Goal and ripple:

$$\sum_{i=0}^{n} \left(\left\lfloor \frac{\boxed{i+1}^{\uparrow}}{\boxed{m+1}^{\uparrow}} \right\rfloor \right)$$

$$\sum_{i=0}^{n} \boxed{\binom{i}{\lfloor m+1 \rfloor} + \binom{i}{\lfloor m \rfloor}}^{\uparrow}$$

$$\boxed{\sum_{i=0}^{n} \binom{i}{\lfloor m+1 \rfloor} + \sum_{i=0}^{n} \binom{i}{\lfloor m \rfloor}}^{\uparrow}$$

Wave-rules:

$$\binom{\boxed{A+1}^{\uparrow}}{\boxed{B+1}^{\uparrow}} \Rightarrow \boxed{\binom{A}{\boxed{B+1}^{\downarrow}} + \binom{A}{B}}^{\uparrow} \tag{5.19}$$

$$\sum_{i=A}^{B} \boxed{U+V}^{\uparrow} \Rightarrow \boxed{\sum_{i=A}^{B} U + \sum_{i=A}^{B} V}^{\uparrow} \tag{5.20}$$

Figure 5.7 Summing series using rippling. The first wave-rule is a simple binomial identity, and the second a distributive law of $\sum$ over $+$. These are applied in turn. The final goal contains two copies of the given and can be fertilized twice.

Given and ripple:

$$true(\;\boxed{a \supset b \land b \supset c}\;^\uparrow\;)$$

$$true(\;\boxed{a \supset b}\;^\uparrow\;) \qquad and \qquad true(\;\boxed{b \supset c}\;^\uparrow\;)$$

$$\boxed{true(a) \;\to\; true(b)}\;^\uparrow \qquad and \qquad \boxed{true(b) \;\to\; true(c)}\;^\uparrow$$

Goal and ripple:

$$true(\;\boxed{a \supset c}\;^\uparrow\;)$$

$$\boxed{true(a) \;\to\; true(c)}\;^\uparrow$$

Wave-rules used:

$$true(\;\boxed{A \supset B}\;^\uparrow\;) \;\Rightarrow\; \boxed{true(A) \;\to\; true(B)}\;^\uparrow$$

$$true(\;\boxed{A \land B}\;^\uparrow\;) \;\Rightarrow\; true(A) \tag{5.21}$$

$$true(\;\boxed{A \land B}\;^\uparrow\;) \;\Rightarrow\; true(B) \tag{5.22}$$

Figure 5.8 Rippling in meta-logical reasoning.

The term true is a meta-logical truth predicate; $\supset$ is implication in the object-logic, and $\to$ is implication in the meta-logic. In Negrete (1994) formulas are also annotated with polarity markers, but these are omitted here for the sake of uniformity. Note that rippling takes place in both the given and the goal.

Given: $(b \sqcap c) \sqcap d = 0$
Additional hypothesis: $a \sqcap (c \sqcup d) = 0$
Goal, ripple and fertilize:

$$\boxed{(a \sqcup \boxed{(b \sqcap c)})}^{\uparrow} \sqcap \boxed{(a \sqcup (b \sqcap d))}^{\uparrow} = a$$

$$a \sqcup \boxed{((b \sqcap c) \sqcap \boxed{(a \sqcup (b \sqcap d))}^{\uparrow})} = a$$

$$...$$

$$a \sqcup \boxed{((b \sqcap c) \sqcap \boxed{(a \sqcap (c \sqcup d)) \sqcup (b \sqcap d)}^{\uparrow})} = a$$

$$a \sqcup \boxed{((b \sqcap c) \sqcap \boxed{(0 \sqcup (b \sqcap d))}^{\uparrow})} = a$$

$$a \sqcup \boxed{((b \sqcap c) \sqcap \boxed{(b \sqcap d)}^{\uparrow})} = a$$

$$a \sqcup \boxed{(b \sqcap ((b \sqcap c) \sqcap d))}^{\uparrow} = a$$

$$a \sqcup \boxed{((b \sqcap c) \sqcap d)}^{\uparrow} = a$$

$$a \sqcup \boxed{0}^{\uparrow} = a$$

$$a = a$$

Wave-rules:

$$\boxed{0 \sqcup X}^{\uparrow} \Rightarrow X$$

$$\boxed{(X \sqcup X)}^{\uparrow} \Rightarrow X$$

$$\boxed{(X \sqcap X)}^{\uparrow} \Rightarrow X$$

$$\boxed{X \sqcap (X \sqcup Y)}^{\uparrow} \Rightarrow X$$

$$\cdots\cdots\cdots$$

$$(X \sqcap Z) = X \rightarrow \boxed{(X \sqcup Y)}^{\uparrow} \sqcap Z \Rightarrow \boxed{X \sqcup (Y \sqcap Z)}^{\uparrow}$$

Figure 5.9 Rippling in SAM's lemma. The given is a part of the hypothesis of the theorem. We use rippling to enable its use on the left-hand side of the goal (red skeleton). After its application, embedding the left-hand side into the right-hand side of the goal results in the blue skeleton and rippling is used again to remove the differences. Rippling is done modulo a built-in associative, commutative idempotent matcher. This makes the ripple steps hard to follow. Above, both rules and goal have been modified slightly to minimize the need for commutativity to make the rippling steps easier to follow.

Givens:
$$size(l) = count(nodes\_in(l))$$
$$size(r) = count(nodes\_in(r))$$

Goal: $size(\ node(n,\ l,\ r)\ ^{\uparrow}\ ) = count(nodes\_in(\ node(n,\ l,\ r)\ ^{\uparrow}\ ))$

Ripple:

$$s(\ size(l) + size(r)\ )^{\uparrow} = count(\ insert(n,\ nodes\_in(l) \cup nodes\_in(r)\ )^{\uparrow}\ )$$

$$s(\ size(l) + size(r)\ )^{\uparrow} = s(\ count(\ nodes\_in(l) \cup nodes\_in(r)\ ^{\uparrow}\ ))^{\uparrow}$$

$$size(l) + size(r)\ ^{\uparrow} = count(\ nodes\_in(l) \cup nodes\_in(r)\ ^{\uparrow}\ )^{\uparrow}$$

$$size(l) + size(r)\ ^{\uparrow} = count(nodes\_in(l)) + count(nodes\_in(r))\ ^{\uparrow}$$

$$size(l) = count(nodes\_in(l)) \wedge size(r) = count(nodes\_in(r))\ ^{\uparrow}$$

Figure 5.10 Rippling in color. This is Figure 2.2 on p40 reproduced using actual colors rather than superscripts. The $\{r, b\}$ label is indicated in purple.

6

From rippling to a general methodology

This book has described rippling: a technique for guiding proof search so that a given may be used to prove a goal. We will investigate in this chapter other areas of automated reasoning involving heuristic restrictions on proof search. We will use problems from these areas to illustrate how the ideas behind rippling can be generalized and used systematically to understand and implement many different kinds of deductive reasoning.

- In many proof calculi, the application of rules in certain situations is known to be unnecessary and can be pruned without sacrificing completeness. For example, in basic ordered paramodulation and basic superposition (Bachmair *et al.*, 1992; Nieuwenhuis & Rubio, 1992), paramodulation is forbidden into terms introduced by applying substitutions in previous inference steps.
- In tactic-based theorem-proving, it is sometimes useful to track parts of the conjecture and use this to restrict proof search. Focus mechanisms (e.g. Robinson & Staples, 1993; Staples, 1995) for this purpose have been developed and hardwired into several calculi.
- In analogical reasoning, a previous proof (the source proof) is abstracted to serve as a proof template for subsequent conjectures (the target conjecture). Additional information about the source proof (in addition to the proof tree) is typically required to compute an abstract proof sketch (Kolbe & Walther, 1994; 1998) for a related target conjecture.

In each of the above techniques, there is a need to encode and maintain information about individual terms and symbols and their inter-relationships. Historically, each technique provided an individual solution to this general problem by introducing a specialized calculus. Inspired by the rippling calculus, we will outline a methodology to augment logic calculi with a generic mechanism to maintain such strategic information. This allows us to describe various control strategies, including those mentioned above, as abstractions on annotated formulas.

The key idea behind the general methodology that we propose is that annotations can be used to distinguish different occurrences of the same symbol. For instance, in rippling the same function symbol may occur in a wave front and a skeleton, and rippling treats these occurrences differently. Abstractions on annotated terms (like the notion of a skeleton in rippling) can map the same term in different ways, depending on how they are annotated.

To realize this key idea, an abstraction on annotated terms denotes a mapping on *occurrences of unannotated terms* rather than a mapping on the *terms* themselves. Suppose we want to abstract two occurrences of the same symbol in a different way. Attaching different annotations to different occurrences of the same symbol, an abstraction (operating on annotated terms) is able to deal with all these occurrences individually. As mentioned in Section 4.2.2, each rippling step amounts to a rewrite step in the corresponding unannotated calculus. Annotations serve to impose additional restrictions on possible rule applications. We are free to use arbitrary information that can be encoded into annotations to restrict the search space. This allows us to formulate many existing proof-search strategies in terms of abstractions on annotated terms. The reason is that, in many cases, the proof search depends on the history or the "semantic context" of individual terms. Annotations provide a technical means to encode such information about a term and associate it with occurrences of individual subterms or even function symbols. By integrating a term and the information about it into a single annotated term, we can manipulate both term and semantic information in a uniform way. Annotated calculi provide inference rules that infer new annotated formulas from existing ones, i.e. they infer new formulas *together* with corresponding deduced semantic information. Therefore, annotations provide a technical means to formalize and implement new proof strategies that rely on context or domain-specific knowledge encoded into annotations, which is automatically propagated during the deduction.

After an introduction to a generalized notion of annotations in Section 6.1, in Section 6.2 we will provide different examples of how to encode strategic knowledge into annotations. Our examples are proof strategies from different areas, starting with the well-known example of rippling, and ending with reuse of proofs based on analogical reasoning. After the discussion of rippling, the examples are selected with respect to the complexity of the annotations used to encode the necessary knowledge. In Section 6.3 we will show how to define appropriate *abstractions* of annotated terms to support the implementation of various proof strategies. To simplify matters, we will use the same examples as presented in Section 6.1. Formalizing the proof strategies with the help of annotations, we will describe the strategies in terms of abstractions on annotated formulas. Rather than presenting isolated solutions to various problems (e.g. basic paramodulation), the examples illustrate the bandwidth of possible applications for using annotations to guide proof search via abstractions. Moreover, together with Section A1.2, our ideas provide the basis of a uniform framework for describing and implementing application-specific proof strategies based on contextual knowledge. We will finish this chapter with a description of different implementations of this methodology.

6.1 A general-purpose annotation language

In order to extend rippling to a general methodology, we start with a discussion of suitable calculi to encode and maintain strategic information. In Section 4.4.1, we contrasted two approaches to annotation: context markers and symbol markers. Although we found it convenient to describe rippling using context markers, the symbol-marking approach is more readily generalized, so we will adopt it here for our general methodology.

What language should be used to mark symbols? In the context-marking approach, we used a finite set of ground annotations (wf_{up}, wf_{down}, wh) to encode the embeddings of the given in the goal. However, when annotating wave-rules, we ran into the problem of weakening wave-fronts, as described in Section 2.4.3. The reason is that the skeleton of a wave-rule encodes the maximal similarity of the left- and right-hand sides. However, the skeleton of a wave-rule may be a proper superset of the skeleton of the goal to which it is applied. In this case, we have to weaken the skeleton of the wave-rule by converting corresponding skeleton parts of left- and right-hand sides to wave-fronts. To facilitate this, we introduce annotation variables, which can be instantiated to either wave-fronts or wave-holes during annotated matching. These annotation variables are used to annotate symbols in wave-rules.

Rippling propagates information about subterm relations between a goal and a given during rewriting. We only have to consider two cases: either a symbol occurrence belongs to the skeleton or it does not. However, in the general case, we may want to encode more complex information. To implement, for example, reuse by analogy, as described in Section 6.2.4, we need to accumulate information about individual symbol occurrences during rewriting. We will need a more expressive annotation language than just a fixed set of annotation constants. Therefore we now introduce the notion of a first-order language parameterized by an annotation signature.

Definition 13 *Let* Π *be an annotation signature and* $\mathcal{V}$ *be a set of annotation variables. Then the term language* $\mathcal{T}_{\Pi(\mathcal{V})}$ *is the set of* annotations.

We use Greek letters such as α, β, and γ to denote annotation variables in $\mathcal{V}$ and use a sans serif font for elements of the annotation signature Π, such as c or d. Using annotations as symbol markers (cf. Section 4) we redefine annotated terms as follows.

Definition 14 *Let* $\mathcal{T}_{\Pi(\mathcal{V})}$ *be a set of annotations,* Σ *be a signature, and* $\mathcal{X}$ *be a set of variables such that* Π*,* $\mathcal{V}$*,* Σ *and* $\mathcal{X}$ *are pairwise disjoint. The set of annotated terms* $\mathcal{A}$ *is the smallest set where*

- $U^{\mathsf{c}} \in \mathcal{A}$ *if* $U \in \mathcal{X}$*,* $\mathsf{c} \in \mathcal{T}_{\Pi(\mathcal{V})}$*, and*
- $f^{\mathsf{d}}(t_1, \ldots, t_n) \in \mathcal{A}$ *if each* $t_i \in \mathcal{A}$*,* $f \in \Sigma$*, and* $\mathsf{d} \in \mathcal{T}_{\Pi(\mathcal{V})}$*.*

Abusing notation, we use $\mathcal{T}_{\Pi(\mathcal{V})}(t)$ to refer to the annotation of the top-level symbol of t, e.g. $\mathcal{T}_{\Pi(\mathcal{V})}(f^{\mathsf{c}}(a^{\mathsf{b}})) = \mathsf{c} = \mathcal{T}_{\Pi(\mathcal{V})}(U^{\mathsf{c}})$. Moreover, we denote the set of all variables of an annotated term t by $\mathcal{X}(t)$.

Analogously to the erasure function based on context markers, we define erasure using symbol markers as follows.

Definition 15 *The* erasure *function* erase: $\mathcal{A} \rightarrow \mathcal{T}_{\Sigma(\mathcal{X})}$ *is defined by* $erase(U^{\mathsf{c}}) = U$ *and* $erase(f^{\mathsf{c}}(t_1, \ldots, t_n)) = f(erase(t_1), \ldots, erase(t_n))$ *for all* $U \in \mathcal{X}$*,* $\mathsf{c} \in \mathcal{T}_{\Pi(\mathcal{V})}$ *and* $f \in \Sigma$*.*

In Section A1.2, we will formalize annotated substitution for symbol markers. For the moment, we require only that an annotated substitution satisfies the following substitutability property.

Lemma 5 (substitutability property) *Let* $\rho : \mathcal{A} \rightarrow \mathcal{A}$ *be an annotated substitution. Then, for all annotated terms* t_1 *and* t_2*, it holds that*

$$erase(t_1) = erase(t_2) \text{ implies } erase(\rho(t_1)) = erase(\rho(t_2)).$$

Analogously to Chapter 4, we define annotated rewriting based on annotated matching and annotated substitution. Let $t \Rightarrow t[\rho(r)]_p$ be an arbitrary rewrite using the rewrite rule $l \Rightarrow r$, where t and $t[\rho(r)]_p$ coincide in all positions that are independent of p and the annotations of corresponding symbol occurrences coincide. The annotated terms t and $t[\rho(r)]_p$ differ only in the subterm at position p, which is the instantiated left-hand side $\rho(l)$ in t and the instantiated right-hand side $\rho(r)$ in $t[\rho(r)]_p$. Analogously, the annotations of subterms t/p and $\rho(r)$ are instantiations of annotations of the corresponding sides of the rewrite rule $l \Rightarrow r$. Thus, the annotations of r determine the annotations of the new subterm $\rho(r)$. Using the same annotation variables on both sides of a rule allows us to transfer information from the replaced subterm to the replacing subterm, since the matcher will instantiate these annotation variables when applying the rewrite rule. Hence, parts of the annotations in t/p, which are matched by an annotation variable in l, will occur again in $\rho(r)$. Hence, also, depending on how we annotate the rewrite rule $l \Rightarrow r$, we are able to propagate different kinds of information during rewriting.

Since we do not need all the technical details of such an annotated calculus in the following sections, we refrain from discussing more details within this section. The reader is referred to Section A1.1 for the complete formal definition of an annotated first-order calculus.

6.2 Encoding constraints in proof search: some examples

To motivate our generalization of rippling, we present various examples where strategic information is used to guide proof search. For each example we show how this information can be maintained during rewriting. We show how different ways of annotating rewrite rules realize different kinds of information propagation. For each of these illustrations we prove the same theorem

$$s(x) + s(y) = s(s(x) + y) \tag{6.1}$$

with the help of the following two rewrite rules

$$s(U) + V \Rightarrow s(U + V) \tag{6.2}$$

$$x + s(V) \Rightarrow s(x + V). \tag{6.3}$$

Here U and V denote meta-variables that can be instantiated by unification, and x denotes a term variable that is treated as a constant (cf. Section 2.2.2).

6.2.1 Example 1: encoding rippling and difference reduction

In this example, we introduce our general-purpose annotation language by considering a familiar application of rippling.

Suppose that formula (6.1) denotes the goal of the inductive theorem $x + s(y) = s(x + y)$, while (6.3) plays the role of the given. Then, using context markers, we would annotate the goal as

$$\boxed{s(x)} + s(y) = s(\boxed{s(x)} + y). \tag{6.4}$$

Since rippling has to preserve the skeleton, we must track each symbol occurrence of the skeleton in each rewrite step. To implement the notions of skeletons and wave-fronts using symbol markers, we annotate the skeleton symbols by s and wave-fronts by wf. Thus, we can present (6.4) using symbol markers as

$$s^{\text{wf}}(x^{\text{s}}) +^{\text{s}} s^{\text{s}}(y^{\text{s}}) = s^{\text{s}}(s^{\text{wf}}(x^{\text{s}}) +^{\text{s}} y^{\text{s}}). \tag{6.5}$$

As rippling is skeleton preserving, applying rewrite rule (6.2) to the left-hand side of (6.5) should result in the formula:

$$s^{\text{wf}}(x^{\text{s}} +^{\text{s}} s^{\text{s}}(y^{\text{s}})) = s^{\text{s}}(s^{\text{wf}}(x^{\text{s}}) +^{\text{s}} y^{\text{s}}). \tag{6.6}$$

To apply rewrite rule (6.2), it must be annotated using symbol markers. Using context markers, we can annotate (6.2) in the following two ways.

$$s(\boxed{U}) \ + V \Rightarrow s(\boxed{U + V}) \tag{6.7}$$

$$\boxed{s(U) + V} \Rightarrow s(\boxed{U + V}) \tag{6.8}$$

How do we get the same effect with symbol markers? Each symbol must be annotated with an annotation variable. To ensure that corresponding symbols are instantiated to an identical marker during rewriting, we must annotate them with the same annotation marker. Otherwise, we use different markers to ensure maximum flexibility. Suppose we use only annotation variables markers, linking the corresponding occurrences of s, $+$, U, and V. This results in

$$s^{\gamma}(U^{\alpha}) +^{\delta} V^{\beta} \Rightarrow s^{\gamma}(U^{\alpha} +^{\delta} V^{\beta}). \tag{6.9}$$

Applying this rule to (6.5) copies the annotation of s in the replaced term to the occurrence of s in the replacing term. However, in general, (6.9) does not preserve the skeleton, as it changes the position of s in the rewritten term. For example, applying wave-rule (6.9) to $s^{\text{s}}(x^{\text{s}}) +^{\text{s}} s^{\text{wf}}(y^{\text{s}})$ produces $s^{\text{s}}(x^{\text{s}} +^{\text{s}} s^{\text{wf}}(y^{\text{s}}))$, which has changed the skeleton from $s(x) + y$ to $s(x + y)$. The solution to this problem is to allow only instantiations of (6.9) that enforce that either s or $+$ is part of a wave-front. Thus, we obtain the following two rules, which replace (6.9):

$$s^{\text{wf}}(U^{\alpha}) +^{\gamma} V^{\beta} \Rightarrow s^{\text{wf}}(U^{\alpha} +^{\gamma} V^{\beta}) \tag{6.10}$$
$$s^{\gamma}(U^{\alpha}) +^{\text{wf}} V^{\beta} \Rightarrow s^{\gamma}(U^{\alpha} +^{\text{wf}} V^{\beta}). \tag{6.11}$$

These correspond to the context-marked wave-rules (6.7) and (6.8), respectively.

Returning to our running example, (6.10) can be used to ripple-out the wave-front on the right-hand side, which results in the annotated formula

$$s^{\text{wf}}(x^{\text{s}} +^{\text{s}} s^{\text{s}}(y^{\text{s}})) = s^{\text{s}}(s^{\text{wf}}(x^{\text{s}} +^{\text{s}} y^{\text{s}})). \tag{6.12}$$

A special rippling rule that simulates the application of annotated tautologies (cf. Section 2.1.4) yields

$$s^{\text{wf}}(x^{\text{s}} +^{\text{s}} s^{\text{s}}(y^{\text{s}})) = s^{\text{wf}}(s^{\text{s}}(x^{\text{s}} +^{\text{s}} y^{\text{s}})), \tag{6.13}$$

which enables (weak) fertilization with the induction hypothesis.

6.2.2 Example 2: encoding basic ordered paramodulation and basic superposition

Paramodulation (Robinson & Wos, 1969) is a theorem-proving method for first-order logic with equality, which is refutationally complete provided various ordering restrictions are imposed. Paramodulation was invented to build in the notion of equality into a resolution-style calculus. Inference rules like resolution, paramodulation, and factorization are used to enlarge a given set of axioms by new inferred formulas until the formula "False" (indicating a contradiction) is deduced. Given two formulas

$$\Psi \rightarrow l = r \tag{6.14}$$

$$\Phi[t], \tag{6.15}$$

paramodulation allows one to infer the formula

$$\sigma(\Psi) \rightarrow \sigma(\Phi[r]) \tag{6.16}$$

provided that σ is the (most general) unifier of t and l. Since derived formulas are added to the set of axioms without removing their parents (the formulas used to infer these formulas), in many cases identical formulas can be derived if we simply permute the inference steps. For instance, suppose there is another equation

$$r = s \tag{6.17}$$

in our set of axioms. Using (6.17) to paramodulate on (6.15) yields

$$\sigma(\Psi) \rightarrow \sigma(\Phi[s]). \tag{6.18}$$

However, we can also use (6.17) to paramodulate on (6.14) to obtain

$$\Psi \rightarrow l = s. \tag{6.19}$$

If we use (6.19) to rewrite (6.15), then we obtain (6.18) again.

Basic superposition (Bachmair *et al.*, 1992; Nieuwenhuis & Rubio, 1992) refines paramodulation in order to reduce such redundancies by forbidding inferences at terms introduced by substitutions from previous inference steps. Basic ordered paramodulation is a further refinement which also forbids inferences at any term positioned below a former paramodulation inference. For example, consider formula (6.16). Since $\sigma(r)$ (occurring inside $\sigma(\Phi(r))$) is located at the position we applied the paramodulation step, we are not allowed to paramodulate further at this position, which prohibits our first way of deducing (6.18). We can only deduce (6.18) by paramodulating on (6.14) and using (6.17) to obtain (6.15). Both refinements are refutationally complete in the context of deletion rules, as explained in Bachmair *et al.* (1992).

To illustrate basic superposition and basic ordered paramodulation, we generalize our running example (6.1) by replacing $s(x)$ with a variable X, yielding

$$X + s(y) = s(X + y). \tag{6.20}$$

Moreover, we will now use narrowing to manipulate this formula. Narrowing generalizes rewriting by replacing matching with unification, so that variables in the goal may be instantiated during inference. Namely, a formula $\rho(t)[\rho(r)]_p$ is deducible from t using narrowing if and only if ρ is a most-general unifier of t/p and the left-hand side l of some rewrite rule $l \Rightarrow r$.

Both basic superposition and basic ordered paramodulation disallow paramodulation into variables, such as X in our example. Superposition of the left-hand side of (6.20) with rewrite rule (6.2) yields

$$s(U + s(y)) = s(s(U) + y). \tag{6.21}$$

Basic superposition forbids inferences at any term introduced as part of the applied substitution. In this case, the substitution is $\{s(U)/X, s(y)/V\}$. Thus, in the rewritten formula, we are not allowed to paramodulate into the term $s(y)$ on the left-hand side or into the term $s(U)$ on the right-hand side. Basic ordered paramodulation introduces further restrictions also forbidding inferences at any term positioned below a former paramodulation inference. In our example, this forbids any inference on the left-hand side of the rewritten formula.

In the original paper (Bachmair *et al.*, 1992) the authors propose an extention of the calculus by introducing Boolean flags attached to each term to implement a marking strategy. Using our general annotation framework we can directly implement these techniques by providing appropriate annotated wave-rules. To do so, we must keep track of information about each subterm, namely whether it was introduced by substitutions from previous inference steps, or whether it is still part of the original conjecture. To achieve this, we introduce two annotation constants b and c: b denotes terms that cannot be paramodulated upon (blocked terms), while c denotes terms that can be paramodulated upon. We allow an inference step on a subterm if and only if its top-level symbol is annotated with c, while an annotation b will block any superposition at this position.

At the start of a proof, only paramodulation into variables is forbidden. Hence all occurrences of variables are labeled with b, while all other symbols are annotated with c. This results in the annotation of conjecture (6.20) as

$$X^b +^c s^c(y^c) =^c s^c(X^b +^c y^c). \tag{6.22}$$

If we superimpose the left-hand side of this conjecture with the annotated version of (6.2), X is instantiated with $s(U)$ and V with $s(y)$. Basic superposition forbids inferences on terms that are introduced as part of an applied substitution. Thus all symbol occurrences in $s(U)$ and $s(y)$ must be annotated with b. Keeping in mind that paramodulation into variables is also forbidden, we require the annotated formula

$$s^c(U^b +^c s^b(y^b)) =^c s^c(s^b(U^b) +^c y^c). \tag{6.23}$$

To automate the process we must account for the following four cases.

(i) Symbol occurrences that descend from the original formula simply inherit their annotations from the corresponding symbol occurrences in the original formula.

(ii) Occurrences of function symbols that are introduced by the right-hand side of a rewrite rule are annotated by c.

(iii) All terms that are introduced by substitutions are annotated by b.

(iv) All occurrences of variables are annotated by b.

Returning to our running example, note that the annotated right-hand side of rewrite rule (6.2) is

$$s^c(U^b +^c V^b). \tag{6.24}$$

Assume that an annotated variable such as X^b or V^b can only be instantiated with a term that is uniformly annotated with b. This seems to be a reasonable condition since substitutions should not affect constant annotations and we must compute annotations for all the individual symbol occurrences of an instantiated term in a deterministic way. Hence, the annotated unifier generated during narrowing is $\{s^b(y^b)/V^b, s^b(U^b)/X^b\}$. The annotated formula (6.23) is uniquely determined by this annotated unifier, the annotated right-hand side (6.24), and the original formula (6.22).

Note that the annotated left-hand side of (6.2) will not contribute to the annotations of (6.23), since whether a subterm in (6.23) is blocked or not does not depend on the blocking information of the replaced term. Hence, we do not care about the annotations of the replaced term except that its top-level symbol is not blocked. We express the "do not care" information by annotating the non top-level positions of the left-hand side with annotation variables. During unification, these annotation variables can be instantiated to arbitrary annotations. Thus, using annotation variables α, β, and γ, we obtain the following annotated version of (6.2):

$$s^{\gamma}(U^{\alpha}) +^{c} V^{\beta} \Rightarrow s^{c}(U^{b} +^{c} V^{b}). \tag{6.25}$$

We summarize this approach to implementing superposition using annotations. The procedure for annotating rewrite rules is defined by the following two annotation rules:

(i) On the left-hand side of a rule, all non-top-level symbol occurrences are annotated by different annotation variables, while the top-level position is labeled with c.

(ii) On the right-hand side, all variable occurrences are labeled with b and function symbols are labeled with c.

It is easy to show that the restrictions on paramodulation imposed by basic superposition are simulated by the restrictions on rewriting imposed by the annotations. Terms that must not be rewritten are annotated by b. Initially, only variables occurring in the conjecture or on the right-hand side of a rule are annotated with b. Hence, any symbol occurrence annotated with b in a rewritten conjecture must be originally introduced by a substitution of such a variable and we only block inferences at terms introduced by substitutions. Since initially all variables are annotated with b, and since a variable X^{b} can only be substituted by a term that is uniquely annotated with b, we inhibit all inference at any subterm introduced by a substitution.

According to our rules for annotating rewrite rules, rewrite rule (6.3) is annotated as

$$x^{\alpha} +^{c} s^{\beta}(V^{\gamma}) \Rightarrow s^{c}(x^{c} +^{c} V^{b}).$$

Basic ordered paramodulation restricts the rewriting further because it also prohibits paramodulation on subterms that originate from the right-hand side of previously used rewrite rules. We simulate this behavior by slightly changing the way of annotating the right-hand side of a rewrite rule. Instead of annotating occurrences of function symbols with c, we mark all symbols on the right-hand side with b. For example, (6.2) would be annotated as

$$s^{\gamma}(U^{\alpha}) +^{c} V^{\beta} \Rightarrow s^{b}(U^{b} +^{b} V^{b}). \tag{6.26}$$

Paramodulating (6.23) with this rule results in the formula

$$s^{b}(U^{b} +^{b} s^{b}(y^{b})) =^{c} s^{c}(s^{b}(U^{b}) +^{c} y^{c}). \tag{6.27}$$

in which all positions on the left-hand side of the formula are now blocked.

Summing up, annotated terms allow us to simulate both basic superposition and basic ordered paramodulation by encoding the given restrictions into the annotations of the initial conjecture and of the given rewrite rules. The rewrite rules as well as the conjecture are annotated in a uniform way, which is easily automated, and in the INKA system (Autexier, 2003) we used a similar approach to implement an efficient simplification procedure that avoids the simplification of already simplified terms.

All non-variable positions of the conjecture are annotated by c, while variable positions are annotated by b. All symbol occurrences of the left-hand side of a rule are annotated by an annotation variable except the top-level position, which is annotated by c. Variables on the right-hand side are annotated by b, while the occurrences of function symbols are annotated by c. In the case of basic ordered paramodulation, the complete right-hand side of a rule is annotated by b. Thus, different restrictions on the inference process are reflected by different annotations of the right-hand sides of the rules.

6.2.3 Example 3: Encoding window inference

Window inference (Robinson & Staples,, 1993; Staples,, 1995) is a commonly used technique for transformational reasoning. It supports the temporary focusing on arbitrary subexpressions of a formula by using decomposition and recomposition rules, which also provide the logical context of the selected subexpressions. This results in a hierarchy of subproblems that co-exist at a single stage of the proof. To implement this technique, the theorem-prover must track the evolution of subexpressions during inference steps. To illustrate the generality of our framework, we will present an implementation of a simplified version of window inference in our annotation framework.

In the previous example, annotations of the replaced subterm do not contribute to the annotations of the replacing subterm. In this example, we show how to propagate information from one to the other. For this and the following examples, we return to a simple rewriting calculus and consider the original problem of proving conjecture (6.1).

First, we define how subexpressions are tracked during rewriting. Let t be a term and p and $\overline{p}$ be positions. Suppose we want to track a subexpression t/p of a term t and use a rewrite rule $l \Rightarrow r$ to replace $t/\overline{p}$ by $\rho(r)$. We distinguish three cases.

(i) If p and $\overline{p}$ denote independent positions in t, then $(t[\rho(r)]_{\overline{p}})/p$ forms the new subexpression to be tracked.

(ii) If $t/\overline{p}$ is a subterm of t/p, then the inference step rewrites the tracked subexpression and again $(t[\rho(r)]_{\overline{p}})/p$ forms the new subexpression to be tracked.

(iii) If t/p is a subterm of $t/\overline{p}$, we distinguish two cases:

   (a) if t/p is some subterm l' of l, then we compute all positions $p_1, \ldots, p_n$ of l' in r and consider $t/(\overline{p}p_1), \ldots, t/(\overline{p}p_n)$ as new subexpressions to be tracked;

   (b) in the second case, t/p is a subterm $\rho(X)/p''$ of some instantiated variable X occurring in l. Then there is a p' such that $p = p'p''$ holds. Let $p_1, \ldots, p_n$ be the positions of X in r, then $t/(p'p_1), \ldots, t/(p'p_n)$ are the new subexpressions to be tracked.

Rather than computing the new foci after each inference step, we propose the use of annotations to transfer foci in each inference step automatically. We will annotate focus terms with f and non-focus terms with u. For example, if we wish to track the occurrence of $s(y)$ on the left-hand side of (6.1) with the help of annotations, then we will annotate the conjecture as

$$s^{\mathsf{u}}(x^{\mathsf{u}}) +^{\mathsf{u}} s^{\mathsf{f}}(y^{\mathsf{f}}) = s^{\mathsf{u}}(s^{\mathsf{u}}(x^{\mathsf{u}}) +^{\mathsf{u}} y^{\mathsf{u}}). \qquad (6.28)$$

After rewriting with an annotated version of (6.2), we require

$$s^{\mathsf{u}}(x^{\mathsf{u}} +^{\mathsf{u}} s^{\mathsf{f}}(y^{\mathsf{f}})) = s^{\mathsf{u}}(s^{\mathsf{u}}(x^{\mathsf{u}}) +^{\mathsf{u}} y^{\mathsf{u}}), \qquad (6.29)$$

in which the focused subterm $s(y)$ in (6.28) is again annotated by f. We obtain this result if we guarantee that the occurrences of U (respectively V) in the annotated rewrite rule are instantiated by the same *annotated* term. We enforce this condition by using the same annotation variable α for both occurrences of U (and β for both occurrences of V), so the rewrite rule (6.2) is annotated as

$$s^{\mathsf{u}}(U^{\alpha}) +^{\mathsf{u}} V^{\beta} \Rightarrow s^{\mathsf{u}}(U^{\alpha} +^{\mathsf{u}} V^{\beta}). \qquad (6.30)$$

Since the matching procedure also matches annotations, (6.30) is not applicable to a goal such as

$$s^{\mathsf{f}}(x^{\mathsf{f}}) +^{\mathsf{f}} s^{\mathsf{f}}(y^{\mathsf{f}}) = s^{\mathsf{u}}(s^{\mathsf{u}}(x^{\mathsf{u}}) +^{\mathsf{u}} y^{\mathsf{u}}) \qquad (6.31)$$

because the annotation of the top-level occurrences of s and $+$ in (6.30) and (6.31) differ. In particular, using u as annotation for s in (6.30) implies that this rule is always applied to a position outside the focus. In order to allow for rewriting in the focus, we would need an additional annotation of rule (6.2) in which u is replaced by f:

$$s^{\mathsf{f}}(U^{\alpha}) +^{\mathsf{f}} V^{\beta} \Rightarrow s^{\mathsf{f}}(U^{\alpha} +^{\mathsf{f}} V^{\beta}). \qquad (6.32)$$

Again, we can make use of annotation variables to merge both variants (6.30) and (6.32) into the single rule

$$s^\gamma(U^\alpha) +^\gamma V^\beta \Rightarrow s^\gamma(U^\alpha +^\gamma V^\beta), \tag{6.33}$$

in which γ can be instantiated arbitrarily either to f or to u as required when applying the rule. However, since γ has to be instantiated in a uniform way, matching the left-hand side of (6.33) to a term such as $s^f(x^f) +^u s^u(y^u)$ fails.

The restrictions that the annotations impose on the matching process guarantee that either a rewrite rule can only be applied in the focus, or a common proper subterm of the left-hand and right-hand sides of the rewrite rule matches the focus. This ensures that the foci (i.e. the subexpressions annotated by f) are always proper terms. This means that once a function symbol is annotated by f, then all symbols occurring in its arguments are also annotated by f.

Summing up, the above example suggests that we can implement window inference using annotations. As in the previous examples, the annotation of the rewrite rules is done in a uniform way and is easily automated. A variant of this approach has been implemented in the INKA system (Autexier, 2003). It is up to the user to determine the initial focus by annotating the conjecture. Common *proper* subterms between the left-hand and right-hand sides of a rule are annotated by annotation variables such that their corresponding symbols share the same annotation variable. Other, non-related symbols are annotated by u.

6.2.4 Example 4: Proving theorems by reuse

Similar conjectures often have similar proofs. In *analogical reasoning* (e.g. Kling (1971); Owen (1990); Melis (1995); Kolbe and Walther (1998)) we exploit this observation to guide the search for a new target proof with the aid of an existing source proof. Annotations can also be used to help in the automation of this proof reuse.

In our setting, a proof is considered to be a sequence of rewrite steps that is determined by the rewrite rules used and the positions at which they have been applied. The proof of the source conjecture is transformed into the starting-point for the target proof. Each of its rewrite steps is translated into the corresponding rewrite rules and positions of the target conjecture (see, for instance, Kolbe and Walther (1994) for details). However, this first approximation to the target proof usually has to be modified by adding or deleting intermediate inference steps. The more sophisticated this modification process is, the more target conjectures can be tackled by reusing a particular source proof. However, inserting or deleting inference steps will hamper the computation of the corresponding positions between the source and target proofs. Adding (or deleting) inference steps in the target proof usually alters the appropriate positions to be used in any subsequent inference steps.

In this section we show how annotations can be employed for reuse. Annotations are used to identify occurrences of subterms in a formula or a proof. The idea is to attach a unique annotation to each symbol occurrence, and to use annotations to identify the positions at which rewrite rules are applied. In Section 6.3.3 we show how this additional information about a (source) proof can be used to guide the (target) proof of a similar conjecture.

In analogical reasoning, an initial mapping is calculated between symbol occurrences in the source and the target conjectures. This has to be extended to a mapping between the source and target proofs because the proofs may contain symbols that do not occur in the conjectures, and because different occurrences of the same symbol may need to be mapped differently if they are treated differently within the source proof. The major advantage of shifting the mapping problem from positions to annotations is that the annotations provide additional information that can guide the automatic extension of the mapping of symbol occurrences between the source and the target proof. To accomplish this, we will relate positions of corresponding proof steps in source and target proofs when they result from a "similar treatment". To identify positions by their treatment, we will store the "proof history" of each symbol occurrence in its annotation.

We use our running example as part of a source proof and track the proof history of each symbol in the proof. To distinguish the different positions in the initial problem (6.1), we annotate each symbol occurrence with a different annotation constant; to simplify notation, we use natural numbers to denote annotation constants:

$$s^2(x^3) +^1 s^4(y^5) = s^6(s^8(x^9) +^7 y^{10}).\qquad(6.34)$$

During rewriting the new subterms must be annotated in such a way that each occurrence has a different annotation. Furthermore, each annotation is entirely determined by the annotations of the initial formulas and by the way the rewrite rules have been applied to the conjecture. If we use a rewrite rule to replace a subterm by a new term, then this new subterm is an instance of the right-hand side of a rule. To distinguish the results of different applications of the same rule, each rule application has to attach new annotations to each symbol of the instantiated right-hand side. Therefore, we will use tuples as the annotations for tracking proof history. These tuples will incorporate information about the applied rewrite rule, the annotation of the position it has been applied to, and the annotation of its original symbol. For instance, the occurrence of s in the right-hand side of the rewrite rule (6.35) is annotated by [a, α, R]: a is a unique annotation constant for this occurrence of s in (6.35), α will be instantiated to the annotation of the position where the rule (6.35) is applied, and R serves as an identifier for the rewrite rule

$$s^\delta(U^\beta) +^\alpha V^\gamma \Rightarrow s^{[a,\alpha,R]}(U^{[\beta,\alpha,R]} +^{[b,\alpha,R]} V^{[\gamma,\alpha,R]}) \qquad (6.35)$$

used in this example. Applying this rule to the left-hand side of (6.34) results in the modified equation

$$s^{[a,1,R]}(x^{[3,1,R]} +^{[b,1,R]} s^{[4,1,R]}(y^{[5,1,R]})) = \ldots . \qquad (6.36)$$

The annotations in (6.36) help us to identify the origin of each symbol occurrence. For instance, x is annotated by [3, 1, R], which denotes that this occurrence of x is transferred from the occurrence 3 of x in the original problem when applying (6.35) to a position denoted by 1. Applying (6.35) to the right-hand side of the equation yields

$$\ldots = s^6(s^{[a,7,R]}(x^{[9,7,R]} +^{[b,7,R]} y^{[10,7,R]})). \qquad (6.37)$$

To collect the proof history in the source proof, we annotate each rewrite rule in the following way: each symbol occurrence on the left-hand side is annotated by a different annotation variable; each occurrence of a function symbol on the right-hand side is labeled by a tuple [s, α, P], where s is a unique annotation constant for this occurrence, α is the annotation variable attached to the top-level symbol on the left-hand side, and P is the identifier for this rewrite rule.[1] For instance,

$$x^\beta +^\alpha s^\delta(V^\gamma) \Rightarrow s^{[c,\alpha,S]}(x^{[e,\alpha,S]} +^{[d,\alpha,S]} V^{[\gamma,\alpha,S]}) \qquad (6.38)$$

is another example of an annotated rewrite rule where S is the identifier of the rewrite rule (6.38). The result of applying this rule to the left-hand side (6.36) of our conjecture is

$$s^{[a,1,R]}(s^{[c,[b,1,R],S]}(x^{[e,[b,1,R],S]} +^{[d,[b,1,R],S]} y^{[[5,1,R],[b,1,R],S]})) = \ldots . \qquad (6.39)$$

Note how the annotations of the symbol occurrences encode the proof history of each symbol occurrence. For instance, y is annotated with [[5, 1, R],[b, 1, R], S], which means that it was inherited from the symbol occurrence labeled with 5 of the original problem by applying the rules (6.35) and (6.38) at the positions 1 and [b, 1, R], respectively.

---

[1] For the sake of readability, we have simplified the annotations: to distinguish different occurrences of the same variable symbol on the right-hand side we must introduce a fourth element to the tuple.

A position in a manipulated formula is (uniquely) characterized by its annotation, which only depends on the annotations of the given conjecture and rules. If we map the annotations of the initial formulas in the source proof to annotations of the initial formulas in the target proof, we are able to identify the corresponding positions of the rewritten source and target conjectures. We can easily automate the annotation of the rewrite rules. All symbol occurrences on the left-hand side of a rule are annotated by different annotation variables, while each symbol occurrence, a, on the right-hand side is annotated by a tuple [c, β, R], where β is the annotation of the top-level symbol of the left-hand side, and R is the unique identifier of the rule. If a is a variable, then c is the annotation of one of its occurrences on the left-hand side, otherwise c is a fresh annotation constant that does not occur elsewhere in the rules or in the conjecture.

The alert reader may wonder about the complexity of annotations, especially for long proofs. Annotations can be efficiently managed by using structure-sharing techniques. For the implementation of such reuse techniques we *reuse* annotations of previous proof parts such that we only have to provide a new instance of a tuple for each symbol occurrence while its elements are annotations of previously existing formulas. These reuse techniques have been implemented in Schairer (1998) as part of a more embracing approach to the general management of change within formal software development.

6.2.5 Summary

In this section we have presented examples of how different kinds of information can be encoded into annotations on formulas. Depending on what information we would like to encode, we adopt different annotations of the given conjecture and the available rewrite rules. As we have seen in Example 1 in Section 6.2.1, we may obtain more than one annotation for each rewrite rule, but in most cases at most one of them is applicable to a specific subterm of the problem. As in the previous examples, the annotation of rewrite rules is done in a uniform way and can be easily automated. Conjectures can often be annotated automatically, but sometimes user interaction is required.

In the next section we will illustrate that all the presented proof strategies can be seen as instances of the proof-by-abstraction paradigm. Annotations will play a vital role to facilitate the formal definition of associated abstractions because they allow us to distinguish different occurrences of the same symbol.

6.3 Using annotations to implement abstractions

According to Giunchiglia and Walsh (1992), abstract proof search is a process by which, starting from a representation of a problem at a so-called ground level, we construct a new and simpler representation at a so-called abstract level, and use this representation to solve the original problem. That is, we abstract the given goal, prove its abstracted version, and then use the information about the resulting abstract proof as an outline to construct the proof at the ground level. In this section, we discuss the benefits of using annotations to define a new family of abstractions. We start by explaining the basic limitations of existing abstraction techniques, and afterward we show how abstractions based on annotation overcome these limitations and open up a new dimension of possibilities. Finally, we demonstrate the flexibility of this approach with the help of the examples introduced in the last section.

Different techniques to abstract-away details have been studied in the literature. One of the main problems is to find out which details should be abstracted-away. On the one hand, if we abstract too much information, then we often obtain abstract solutions that cannot be transferred to the ground level. Moreover, planning at the abstract level may be even more difficult than planning at the ground level because the abstraction removes useful "control" information. On the other hand, if we abstract too little, then the complexity of finding a proof at the abstract level may be just as hard as at the ground level.

To guide proof search, we demand that abstractions preserve provability; i.e., if there is a proof at ground level, then there is also a corresponding proof at the abstract level. Abstractions that preserve provability are called *PI-abstractions*.[1] Examples of PI-abstractions on standard (i.e. unannotated) terms are, for instance, to identify constants, functions, or predicate symbols Hobbs (1985); Plaisted (1980); Giunchiglia & Walsh (1992) or to drop arguments to function or predicate symbols (Melham, 1990).

6.3.1 Limitations of abstractions

In order to illustrate the benefits of annotations for defining abstractions, we discuss the limitations of standard abstraction techniques on unannotated terms. Experience has shown that abstractions of unannotated terms are of limited use (Plaisted, 1980). In particular, the preconditions of PI-abstractions impose severe restrictions on possible abstractions. We illustrate this below with the help of rewrite systems.

---

[1] PI is an abbreviation for proof invariant, e.g. the abstraction of a proof is a proof at the abstract level.

Let *abs* be an abstraction mapping that maps terms at the ground level to terms at the abstract level. Then, let R be the rewrite system at the ground level and R' be another rewrite system at the abstract level consisting of all abstracted rewrite rules of R that are not tautologies. This means that R' is defined by $R' = \{abs(l) \Rightarrow abs(r) \mid l \Rightarrow r \in R \text{ and } abs(l) \neq abs(r)\}$. A rule $l \Rightarrow r$ with $abs(l) = abs(r)$ can be omitted since the resulting abstracted rule would be a tautology.

We may ask what are the necessary conditions for *abs* to be a PI-abstraction? Suppose *abs* is a PI-abstraction. Then each rewrite step[1] $s \Rightarrow_{l \Rightarrow r} t$ at the ground level can be simulated at the abstract level by a sequence of rewrite steps $abs(s) \Rightarrow^* abs(t)$. If $abs(s) = abs(t)$, then this property holds trivially, so we assume $abs(s) \neq abs(t)$. Suppose, as an additional precondition, that we wish to use the abstracted rewrite rule $abs(l) \Rightarrow abs(r)$ to simulate the inference step $s \Rightarrow_{l \Rightarrow r} t$ at the abstract level. Hence, there should be some position p' in $abs(s)$ and some substitution σ' such that the following holds:

$$s/p = \sigma(l) \text{ implies } abs(s)/p' = \sigma'(abs(l)) \tag{6.40}$$

$$abs(s[\sigma(r)]_p) = abs(t) = abs(s)[\sigma'(abs(r))]_{p'}. \tag{6.41}$$

If we further assume that the top-level position ϵ at the ground level corresponds to the top-level position ϵ at the abstract level, then (6.40) can be simplified to the property that substitutions and the abstraction mapping *abs* commute:

$$abs(\sigma(l)) = \sigma'(abs(l)). \tag{6.42}$$

As a consequence, (6.41) can be simplified to the property that subterm replacement and the abstraction mapping *abs* commute:

$$abs(s[w]_p) = abs(s)[abs(w)]_{p'}. \tag{6.43}$$

Thus, in order to be a PI-abstraction, an abstraction *abs* has to fulfill the constraints (6.42) and (6.43), unless we drop one of our two preconditions. The first precondition requires that the abstract inference step uses the abstraction of the rewrite rule. Dropping this condition would hamper the speculation of suitable rewrite rules at the ground level with the help of the abstract proof. The second precondition demands that the top-level position of a ground term corresponds to the top-level position in the abstraction, which is satisfied by almost all known abstractions.

---

[1] We use the subscript $l \Rightarrow r$ to indicate the rewrite rule $l \Rightarrow r$ used to perform the rewrite step.

Summing up, (6.42) and (6.43) are true in all PI-abstractions of practical use. Roughly speaking, we are only free to specify how abstractions behave on signatures, which then determines how abstractions extend to terms, substitutions, and positions. Hence, we usually end up with abstractions that either identify signature elements, or change their arity. But in practice these kinds of abstractions are too coarse to allow, for example, hierarchical proof search (cf. Plaisted (1980)). For instance, consider the proof strategies we presented before. In rippling, we need additional information to distinguish wave-fronts and skeletons in order to compute abstractions of formulas that are based on the positions of the wave-fronts relative to the skeleton. In the reuse example, additional information about the history of individual terms is needed to define suitable abstractions expressing the abstract structure of the proof.

6.3.2 Abstractions on annotated terms

The use of annotation facilitates a new dimension of possible abstractions. Instead of mapping the signature at the ground level to a signature at the abstract level, we define abstractions on annotated terms as mappings that operate on *tuples*, each containing a signature and annotations. This allows us to distinguish different symbol occurrences (labeled with different annotations) and abstract them in different ways.

Consider the formula (6.4) in the rippling example of Section 6.4. Attaching different annotations to the occurrences of s in (6.4) allows us to map these occurrences either to an abstract wave-front or to a part of the skeleton. Moving from abstractions on unannotated terms to abstractions on annotated terms results in an implicit generalization that allows us to deal with the coincidence of instantiations of different abstract symbols at the ground level. Thus, abstractions of annotated terms denote mappings on *symbol occurrences* rather than mappings on *symbols*. By attaching different annotations to different occurrences of the same symbol, an abstraction may treat these occurrences individually. In our example, we can annotate the first occurrence of s with F, and the second occurrence with G, and then define an abstraction by mapping s^{F} to F and s^{G} to G. Given a term and its desired abstraction, we can incorporate all necessary additional information about this abstraction into annotations. Then, a term, together with its abstraction, is incorporated into a single *annotated* term.

The benefits of this approach are obvious. The annotated calculus will automatically propagate the annotations when doing an inference step. Hence, this inference step will produce the deduced formula *together* with its abstraction. Analogously, annotated rules are used to combine rules and their abstractions.

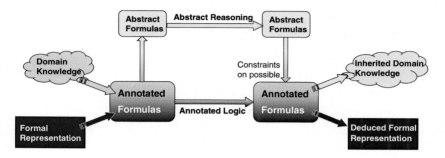

Figure 6.1 Abstractions on annotated formulas.

During proof search we can select appropriate rules according to their abstractions and – knowing the inference mechanisms of the annotated calculus – anticipate the resulting abstraction of the deduced formula. Figure 6.1 illustrates this approach.

Within this approach, we are free to encode arbitrary knowledge into annotations provided that the inference mechanism of annotations "preserves" the semantics of the encoded knowledge. This allows us to encode many existing proof strategies in terms of annotated terms that take into account the history or semantic background of individual terms. Annotations provide the technical means to incorporate such information about a term into its own representation. The annotated calculus manipulates both terms and the information about them in a uniform way.

As mentioned before, control information is encoded into annotations that enable us to define abstractions that use strategic information from subformulas. While an abstraction may remove information about the term syntax, it can propagate the strategic information encoded in the annotations to the abstract level.

The definition of appropriate abstractions depends on the strategic information that we want to incorporate into annotations. In general, the search at the ground level can be guided by predicting some properties of the intermediate steps of the rewriting. In equality reasoning, for instance, we aim to rewrite a given problem into a trivial equation $l = l$. Techniques have been developed that guide the proof by reducing syntactic differences (Digricoli (1980); Quinlan & Hunt (1986); Autexier & Hutter (1997)). Thus, syntactic similarities must be preserved during deduction.

In general, the given proof information is too weak to predict an intermediate result of the rewriting process precisely, but we often know some properties of it. For instance, manipulating the induction step should result in a formula in which the induction hypothesis occurs as a subformula. We do not know in advance the shape of the wave-fronts in the rewritten formula, but we do know at which positions they may occur. Our aim is to use such information to speculate properties of intermediate steps in the rewriting process.

There are different approaches to using abstractions for guiding proof search depending on the search space at the abstract level. As we show in the examples that follow, there are abstractions that result in search spaces without any branching or with only a finite number of branches. For reuse purposes, the abstraction of the source proof serves also as the abstraction of the target proof. In such cases, we do not need to reason formally at the abstract level; instead, it is sufficient to enumerate the finitely many solutions.

If we use a formal system, such as a rewrite system, to reason at the abstract level, then we must take care of the properties mentioned in Section 6.3.1. The abstraction of an annotated rewrite system is basically determined by the abstraction of annotated symbols, since it has to satisfy the properties (6.42) and (6.43). But, in contrast to the unannotated case, we are now able to distinguish between different occurrences of the same symbol by attaching different annotations to them. Thus, abstractions in the annotated calculus represent abstractions on symbol occurrences in the unannotated calculus. The examples given in Section 6.2 illustrate the wide scope and variety of these abstractions.

Once we find a proof at the abstract level, each abstract proof step has to be refined to a sequence of proof steps at the ground level. Suppose there is a rewrite step $s' \Rightarrow_{abs(l) \Rightarrow abs(r)} t'$ at the abstract level, then we must find a sequence $s = s_0 \Rightarrow \ldots \Rightarrow s_{n-1} \Rightarrow_{l \rightarrow r} s_n = t$ at the ground level such that $s' = abs(s)$ and $t' = abs(t)$. The rewriting $s_0 \Rightarrow \ldots \Rightarrow s_{n-1}$ denotes the preparatory steps that are required to enable the application of $l \Rightarrow r$. During these preparatory steps, the abstractions of the intermediate results should be invariant, i.e. $abs(s_0) = abs(s_i)$, for all i, $1 \leq i \leq n-1$. This information can be used to restrict possible rewrite steps in this phase. Thus, we must identify those rewrite rules whose application is *abs*-invariant. It is easy to prove that the properties (6.42) and (6.43), stated in Section 6.3.1, are sufficient to guarantee that all rewrite rules $l \Rightarrow r$ with $abs(l) = abs(r)$ possess this property.

Following the description above, we obtain a general procedure to instantiate abstract proofs. Each abstract rewrite step suggests the application of a corresponding rewrite rule at the ground level. In order to enable its application, we must first perform some preparatory rewrite steps that are *abs*invariant.

6.3.3 Examples revisited

To illustrate the flexibility of the approach presented above, we return to the examples given in Section 6.2. We start with the simplest example of implementing basic paramodulation, and end with the more complex example of reusing former proofs for proof search. For each example, we show how strategic information can be preserved when mapping the problem to an abstract level.

Basic paramodulation

The simplest way to use annotations is to restrict possible proof steps by using annotations to encode constraints. Similar to standard unification, an annotated unifier identifies two annotated terms, including their annotations. Non-unifiable annotations result in a clash during the unification and thus prevent the application of a rewrite rule. Given an annotated matcher σ of two annotated terms s and t (i.e. $\sigma(s) = t$) then $erase(\sigma)$ is a matcher of the terms $erase(s)$ and $erase(t)$ (i.e. $erase(\sigma)(erase(s)) = erase(t)$) but not vice versa.

In the implementation of basic paramodulation in Section 6.2.2, for instance, we use the annotation **b** to denote blocked terms. Annotating the top-level position of the left-hand side of a rewrite rule by **c** causes a clash when matching this term with a subterm labeled with **b**. This restricts the application of such rules to non-blocked terms. This is similar to rippling, where rewriting may change wave-fronts but not the skeleton. From this point of view, basic paramodulation is a special case of rippling and we can use the same kind of abstractions to implement both.

Rippling

Rippling provides a hierarchy of different abstractions, depending on how much detail we wish to be available at the abstract level. The most abstract level is obtained by using the skeleton as an abstraction. The corresponding abstraction maps each annotated term to its skeleton (or, more generally, the set of its skeletons) as illustrated in Section 6.2.1. For instance, we obtain the set

$$\{x + s(y) = s(x + y)\} \tag{6.44}$$

as an abstraction of conjecture (6.5). This set is also an abstraction of the rewritten conjecture (6.6). Since rippling requires that each rewrite step preserves the skeleton, the abstraction of a rippling proof results in a trivial "proof" in which no rewrite steps are made at the abstract level. While the abstraction restricts possible rewrite steps at the ground level to those steps preserving the skeleton, it provides no information about how to select a suitable rewrite rule to ripple wave-fronts into specific positions. Each rewrite step at the ground level is part of a "preparation" phase that is invariant with respect to *abs*. Since the notion of skeleton satisfies the properties (6.42) and (6.43), we can guarantee that applying skeleton-invariant rewrite rules (i.e. wave-rules) always results in skeleton-invariant rewriting.

To encode more control information into annotated terms, we present an abstraction that preserves the location of wave-fronts. This abstraction maps each wave-front to a unary function •. In our example, we abstract (6.5) to the singleton

$$\{\bullet(x) + s(y) = s(\bullet(x) + y)\}, \tag{6.45}$$

while (6.6) is abstracted to

$$\{\bullet(x + s(y)) = s(\bullet(x) + y)\}. \tag{6.46}$$

Using this abstraction we are able to describe a rippling-out proof for a conjecture $P(f_1(\ldots f_n(\bullet(x))))$ in an abstract way, namely

$$P(f_1(\ldots f_n(\bullet(x)))) \Rightarrow P(f_1(\ldots \bullet(f_n(x)))) \Rightarrow \ldots \Rightarrow \bullet(P(f_1(\ldots f_n(x)))).$$

The • abstraction violates the commutativity property formulated in (6.43), since it is not compatible with term constructors. For example, $f^s(s^{wf}(a^s))$ is mapped to the abstracted term $f(\bullet(a))$. But replacing a^s by $s^{wf}(b^s)$, which corresponds to the abstract rule $a \Rightarrow \bullet(b)$, results in the term $f^s(s^{wf}(s^{wf}(b^s)))$ with an abstraction $f(\bullet(b))$. Informally speaking, the abstraction provides no means to reason about the merging or splitting of wave-fronts. Thus, this abstraction is unsuitable for planning at an abstract level with the help of a rewrite system. However, for a given problem, the search space at the abstract level is finite, since there are only finitely many positions at which • may occur. Thus, instead of planning abstract proofs, we may enumerate different abstract solutions by a search algorithm. In general, more than one rewrite step is required to instantiate an abstract step. The abstractions of the intermediate results may not fit into the abstract proof, as a wave-front may partly already be moved to the new position but may partly remain at the old position. But when repeatedly applying rules that move wave-fronts in the right directions, we will eventually achieve our goal.

We will next consider a more fine-grained abstraction that keeps track of the width of wave-fronts as introduced in Section 4.7.1. We abstract wave-fronts by the number of •, but in this case the abstraction is obtained by replacing *each* occurrence of a function between the root of the wave-front and the wave-hole by an occurrence of •. We can nest arbitrarily many occurrences of •, such as $\bullet(a), \bullet(\bullet(a)), \bullet(\bullet(\bullet(a)))$, etc. While the abstraction of (6.5) is still equal to

(6.45), the equation

$$half^{\text{wf}}(s^{\text{s}}(s^{\text{s}}(U^{\alpha}))) = s^{\text{s}}(half^{\text{wf}}(U^{\alpha})) \qquad (6.47)$$

has the following abstraction in our new notion:

$$\{half(\bullet(\bullet(U))) = \bullet(half(U))\}. \qquad (6.48)$$

Reasoning about the width of wave-fronts at the abstract level, we lose the nice property that the search space at the abstract level is finite. But this abstraction allows us to model appropriately the replacement of a^{s} by $s^{\text{wf}}(b^{\text{s}})$ in a term $f^{\text{s}}(s^{\text{wf}}(a^{\text{s}}))$ at the abstract level by replacing a by $\bullet(b)$ in $f(\bullet(a))$, which results in an abstract term $f(\bullet(\bullet(b)))$. As illustrated in Section 4.6.5, this allows for the definition of termination orderings for annotated rewriting. Moreover, provided we define an appropriate rewrite relation at the abstract level, the corresponding definition of this abstraction will satisfy the conditions (6.42) and (6.43) on PI-abstractions.

Window inference

In Section 6.2.3, we introduced the notion of focus of attention to track occurrences of subterms during rewriting. We use this to track individual subterms during deduction in order to implement a hierarchical version of a difference-reduction approach to solving equations. The idea is to identify common subterms on both sides of the equation under consideration, and to focus rewriting on the remaining differences. During deduction we use the technique of focusing subterms to track the positions of identical subterms on both sides and try to minimize the differences of the context in which the corresponding subterms occur. The context of a subterm is given by the sequence of function symbols (plus the argument position to descend to the subterm) occurring on the path between the root of the term and the subterm. We illustrate this below, returning to the example used throughout Section 6.2.

Suppose we wish to prove (6.1) using difference-reduction techniques. We will assume that the two occurrences of y in (6.1) are related. We focus on both occurrences of y and obtain the annotated equation

$$s^{\text{c}}(x^{\text{c}}) +^{\text{c}} s^{\text{c}}(y^{\text{b}}) = s^{\text{c}}(s^{\text{c}}(x^{\text{c}}) +^{\text{c}} y^{\text{b}}). \qquad (6.49)$$

To track the symbol occurrences, we use the following annotated versions of (6.2) and (6.3):

$$s^{\gamma}(U^{\alpha}) +^{\gamma} V^{\beta} \Rightarrow s^{\gamma}(U^{\alpha} +^{\gamma} V^{\beta}) \qquad (6.50)$$

$$x^{\alpha} +^{\beta} s^{\rho}(V^{\gamma}) \Rightarrow s^{\rho}(x^{\alpha} +^{\beta} V^{\gamma}). \qquad (6.51)$$

To prove (6.49) using difference reduction techniques, we have to rewrite the two sides of the equation into identical terms. Suppose y occurs in each of the subsequent inference steps on both sides of the manipulated equation. Since, at the end of this process both sides of the equation coincide, the contexts in which y occurs will be identical on both sides. Thus, rewriting will equate the positions of y on both sides and also equate the paths of function symbols that occur on the way from the top level to the positions of y. Initially, y occurs in the second argument of $+$ and in the first (and only) argument of s on the left-hand side, which we denote by a list $\langle +_2, s_1 \rangle$. On the right-hand side we obtain $\langle s_1, +_2 \rangle$ as a path description to y. To plan the proof at the abstract level we only consider these paths to y and try to equate these paths. For this reason we also abstract the given rewrite rules (6.50) and (6.51).

For each rewrite rule and for each subterm that matches with y and occurs on both sides, we obtain an abstract rewrite rule. In the case of the rewrite rules (6.50) and (6.51), we obtain the following abstract rewrite rules:

$$\langle +_1, s_1 \rangle \Rightarrow \langle s_1, +_1 \rangle \text{ from (6.50) and } U \tag{6.52}$$

$$\langle +_2 \rangle \Rightarrow \langle s_1, +_2 \rangle \text{ from (6.50) and } V \tag{6.53}$$

$$\langle +_2, s_1 \rangle \Rightarrow \langle s_1, +_2 \rangle \text{ from (6.51) and } V. \tag{6.54}$$

A path to a specific symbol occurrence is represented as a list over $\Sigma \times \mathcal{N}$. Rules (6.52)–(6.54) represent rewrite rules operating on paths, which are abstractions of the context in which a specific symbol occurrence is located. Using these rules for rewriting allows us to replace segments of these paths. Since we use a first-order logic on the ground level, on the abstract level there are no variables occurring in the abstract rules. The left-hand and right-hand sides of an abstract rule denote paths to symbol occurrences, which consist only of function symbols, since first-order logic does not provide any (higher-order) function variables.

In the given example, we obtain the following abstract proof:

$$\langle +_2, s_1 \rangle = \langle s_1, +_2 \rangle \tag{6.55}$$

$$\Rightarrow {}_{(6.53)} \langle s_1, +_2, s_1 \rangle = \langle s_1, +_2 \rangle \tag{6.56}$$

$$\Rightarrow {}_{(6.54)} \langle s_1, s_1, +_2 \rangle = \langle s_1, +_2 \rangle \tag{6.57}$$

$$\Rightarrow {}_{(6.53)} \langle s_1, s_1, +_2 \rangle = \langle s_1, s_1, +_2 \rangle. \tag{6.58}$$

We find such proofs by implementing difference reduction techniques on lists (Autexier & Hutter, 1997). The difference between two lists triggers the use of suitable abstracted rewrite rules.

The abstract proof provides a proof sketch for the proof at the ground level. In our example, we already obtain the ground-level proof if we simply apply the rewrite rules corresponding to those used at the abstract level in an analogous way. We do not have to insert additional preparatory steps.

In general, however, we must search for appropriate preparatory steps that enable the use of the speculated rewrite rules. Since a preparatory step may not change the abstraction of the actual problem, it must keep the path to the focused subterms (such as y in our example) unchanged. This property can be guaranteed in the following way. First, arbitrary rewrite rules can be applied only at positions that are independent of the position of the focus. Second, to manipulate a subterm containing parts of the path, we need appropriate annotated rewrite rules that enforce that the path will not change during a preparatory step. For example,

$$(X^c +^c (Y^\alpha +^c Z^c)) \Rightarrow (Z^c +^c (Y^\alpha +^c X^c))$$

illustrates the typical form of such a rewrite rule. Corresponding symbol occurrences, which posses the same path from the top-level to their positions, are annotated with common annotation variables, while all other symbol occurrences are labeled with c. Using such rules will automatically enforce that the paths of focused expressions will not change during rewriting.

Reuse of proofs

In proof reuse, an abstract proof for the target conjecture is obtained by abstraction of the source proof. In contrast to rippling, it is not necessary to search for a proof at the abstract level. The source proofs instead provide a collection of abstract proofs that can be reused in subsequent target proofs. The more we abstract from the source proof, the more general the abstract proof will be, but the more work we have to do to fit the abstract proof to the target conjecture.

In our example we show how this can work using a simple approach: since we can distinguish between different symbol occurrences by inspecting their annotations, we eliminate the function names in the abstractions.

Consider the reuse example in Section 6.2.4. Abstraction from the symbol names results in the following abstract rewrite rules:

$$\bullet^\alpha(\bullet^\delta(U^\beta), V^\gamma) \Rightarrow \bullet^{[a,\alpha,R]}(\bullet^{[b,\alpha,R]}(U^{[\beta,\alpha,R]}, V^{[\gamma,\alpha,R]})) \qquad (6.59)$$

$$\bullet^\alpha(\bullet^\beta, \bullet^\delta(V^\gamma)) \Rightarrow \bullet^{[c,\alpha,S]}(\bullet^{[d,\alpha,S]}(\bullet^{[e,\alpha,S]}, V^{[\gamma,\alpha,S]})). \qquad (6.60)$$

Also the source proof is abstracted, which yields the following abstract proof:

$$\bullet^1(\bullet^2(\bullet^3), \bullet^4(\bullet^5)) = \dots \qquad (6.61)$$

$$\bullet^{[a,1,R]}(\bullet^{[b,1,R]}(\bullet^{[3,1,R]}, \bullet^{[4,1,R]}(\bullet^{[5,1,R]}))) = \dots \qquad (6.62)$$

$$\bullet^{[a,1,R]}(\bullet^{[c,[b,1,R],S]}(\bullet^{[d,[b,1,R],S]}(\bullet^{[e,[b,1,R],S]}, \bullet^{[[5,1,R],[b,1,R],S]}))) = \dots. \qquad (6.63)$$

Below we reuse the original proof in Section 6.2.4 to guide an inductive proof of the theorem

$$len(x <> (u :: y)) = s(len(x <> y)),\qquad(6.64)$$

where $::$ is the usual constructor for lists, $<>$ denotes the concatenation of lists, and *len* computes the length of a list. To emphasize the similarity of the proofs, we will use prefix notation for $<>$ and $::$ in the following. Hence, we rephrase our problem as

$$len(<>(x, ::(u, y))) = s(len(<>(x, y))).\qquad(6.65)$$

In a first step, we map the symbol occurrences of the source problem to symbol occurrences of the target problem. As mentioned in Section 6.2.4, this is done by labeling related occurrences with identical annotations. Notice that there is no unique solution to this problem. Depending on which parts of the source or target problem are regarded as "similar", we obtain different solutions. For lack of space, we do not go into the details of how such mappings are computed; the reader can consult the literature on analogy and reuse, e.g. Owen (1990); Kolbe and Walther (1994).

Analogously to (6.61), we annotate the given induction conclusion

$$len^r(<>^1(::^2(v^r, x^3), ::^4(u^r, y^5))) = \dots,\qquad(6.66)$$

so that corresponding symbol occurrences of the source and target induction conclusions share common annotations. We use r to annotate symbol occurrences that do not correspond to occurrences in the source proof. Suppose the definitions of *len* and $<>$ provide the following rewrite rules:

$$<>(::(X, U), V) \Rightarrow ::(X, <>(U, V))\qquad(6.67)$$

$$len(<>(x, ::(W, V))) \Rightarrow s(len(<>(x, V))).\qquad(6.68)$$

Corresponding to the rewrite rules (6.59) and (6.60) in the source problem, we obtain annotated versions (6.69) and (6.70) of the rewrite rules above.[1]

$$<>^\alpha(::^\delta(X^\rho, U^\beta), V^\gamma)$$
$$\Rightarrow ::^{[a,\alpha,R]}(X^{[r,\alpha,R]}, <>^{[b,\alpha,R]}(U^{[\beta,\alpha,R]}, V^{[\gamma,\alpha,R]}))\qquad(6.69)$$
$$len^\rho(<>^\alpha(x^\beta, ::^\delta(W^\tau, V^\gamma)))$$
$$\Rightarrow s^{[c,\alpha,S]}(len^{[r,\alpha,S]}(<>^{[d,\alpha,S]}(x^{[e,\alpha,S]}, V^{[\gamma,\alpha,S]})))\qquad(6.70)$$

The annotations of (6.70) reflect a non-trivial mapping from the source to the target problem. Since $<>$ is mapped to $+$, the top-level symbol *len* has no

---

[1] Notice that the symbol mapping, which translates the source problem to the target problem, must be consistent with the translation of annotated source rules to annotated target rules.

counterpart in the source proof and is annotated with $[r, \alpha, S]$ on the right-hand side.

In the first step of the abstract proof we applied (6.59) at the position labeled with 1. Analogously we apply (6.69) to (6.66) at the corresponding position annotated with 1. This results in the formula

$$len^r(::^{[a,1,R]}(v^{[r,1,R]}, <>^{[b,1,R]}(x^{[3,1,R]}, ::^{[4,1,R]}(u^{[r,1,R]}, y^{[5,1,R]}))))= \ldots.$$

Symbol occurrences that are annotated with terms containing r are not related to the source proof symbols. Ignoring these symbol occurrences we obtain (6.62) as the corresponding abstraction. The next source step uses (6.38) and thus suggests the use of (6.60) at the position labeled with $[b, 1, R]$. Under the mapping of (6.38) to (6.70), we must apply (6.70) at the top-level. But, unfortunately, the rule is not applicable because *cons* occurs inside of *len*. Thus, we first carry out a preparatory step that enables the application of (6.70). In particular, we use a rewrite rule from the definition of *len*. To preserve the mapping between symbol occurrences in the source and target proof, we apply an annotated version of this rule, which is able to inherit the necessary information:

$$len^\alpha(::^\beta(X^\gamma, Y^\delta)) \Rightarrow s^\beta(len^\alpha(Y^\delta)). \tag{6.71}$$

This rewrite rule is annotated according to the rules we described in Section 6.2.1. Roughly speaking, rippling tracks the movement of symbol occurrences during an inference step and this is exactly what we want to do when constructing proofs from first principles. We have to transfer the reuse information until we are able to resume the reuse process. Applying this "rippling" rewrite rule at top-level on the left-hand side yields

$$s^{[a,1,R]}(len^r(<>^{[b,1,R]}(x^{[e,1,R]}, ::^{[4,1,R]}(u^{[r,1,R]}, y^{[5,1,R]}))))= \ldots. \tag{6.72}$$

Notice how the use of such generated rewrite rules eases the maintenance of the symbol mapping when patching the source proof. After each step during the patch, the annotations provide the appropriate information necessary to determine the positions at which to apply (6.70) and other rewrite rules. Hence, we apply (6.70) at the corresponding position to obtain

$$s^{[a,1,R]}(s^{[c,[b,1,R],S]}(len^{[r,[b,1,R],S]}(<>^{[d,[b,1,R],S]}(x^{[e,[b,1,R],S]},$$
$$y^{[[5,1,R],[b,1,R],S]}))))= \ldots.$$

In reuse, we use annotations to maintain the mapping between symbol occurrences in the source and target proof: symbol occurrences are related if they share the same annotations. Using terms as annotations, we are able to encode sufficient information about the proof history into annotations to allow for more general versions of analogy. Since annotated rewrite rules, as they are presented in Section 6.2.1, inherit such information, we are able to maintain the mapping also when patching the original proof.

Since our main focus is the general methodology behind rippling, we refrain from a more detailed analysis of this reuse technique, but we would like to emphasize that, by encoding the proof history into annotations, we are able to treat different proof histories in the same way by imposing an equational theory on annotations. For instance, we can ignore the applications of a rewrite rule R by applying an "annotation" rewrite rule $[\alpha, \beta, R] \rightarrow \alpha$ to modify the annotations occurring in the abstract proof. If the application of two rewrite rules is permutative, we can ignore the sequence in which rewrite rules are applied by using some kind of distributivity law on annotations.

6.4 Implementation

The different strategies discussed in this chapter have all been implemented and tested within a common framework, and are described in detail in Autexier *et al.* (1999); (2000); (2002). Initially, these different applications of annotations were individually encoded into separate deduction tools. Later, Serge Autexier designed and implemented a general framework (see Autexier 2003) for hierarchical contextual reasoning. This framework became the common underlying proof engine CoRe of the MAYA (Autexier *et al.*, 2002; Autexier & Hutter, 2002) and OMEGA (Benzmüller *et al.*, 1997) system in 2002. CoRe uses an annotated higher-order calculus (Hutter & Kohlhase, 2000) based on the typed λ-calculus. The supported language of annotations is similar to the language of annotations that we use to define an annotated first-order calculus in Section A1.2. CoRe supports hierarchical planning by providing hierarchical proof data-structures as first-class citizens. To implement different proof strategies (such as those mentioned above), a user has only to provide the following information to CoRe:

(i) the rules defining how to annotate the axiomatization of a problem, which also implicitly pre-determines how information is maintained during proof search;

(ii) the abstraction function;

(iii) tactics to implement a search engine operating on the abstract level and generating abstract proofs; and

(iv) tactics to refine an individual abstract proof step into a sequence of calculus steps at the ground level.

Using CoRe, we do not have to modify the logic engine to make use of domain knowledge (as was done in original approaches of implementing reuse, window inferencing and basic paramodulation). Instead, we encode the required extra knowledge into formula and rule annotations. Tactics operating on annotated formulas (or their abstractions) can make use of this knowledge during deduction. Annotations allow the user to include application-specific domain knowledge into proof search without changing the underlying proof systems. Regardless of how tactics or annotations are defined, the soundness of the underlying calculus is always guaranteed because each proof step is justified in terms of an inference rule in the underlying (presumed sound) calculus.

6.5 Summary

In this chapter, we have generalized the rippling methodology by presenting a framework for embedding and maintaining different kinds of control information using annotations. Based on these annotations, we defined various abstractions in order to simulate different kinds of proof strategies within the proof-by-abstraction paradigm. Considering these different proof strategies as instances of a common methodology has allowed us to implement them in a uniform way using a generic framework for proof search based on abstractions on annotated terms. Diverse illustrative examples were chosen to give the reader a feeling for the scope of this general methodology.

The main application of our work is to formal methods of system development, so our evaluation has been chiefly carried out on the proof obligations arising from formal verification. During this work, we have discovered a number of exciting and productive applications of annotated rewriting, which improve the automated support of formal methods. Here are two such applications.

(i) Proof obligations often contain structural information that may be lost when encoding them in a first-order or higher-order logic. For instance, each part of a proof obligation corresponds to part of the initial specification. Knowing this correspondence eases the selection of an appropriate subformula with which to start the proof search. We use annotations to implement this kind of origin-tracking.

(ii) Analyzing the time spent in carrying out large verifications has revealed that about 50% of the overall time spent on software verification involves repairing or redoing previously correct proofs that have became invalid due to changes in the specification. Appropriate support for the reuse of proofs is essential when transferring formal methods into industry. Again, annotations have been successfully used to support the reuse the proofs in our tools (Melis & Schairer (1998); Schairer (1998)).

Annotated rewriting has provided powerful and discriminating tools for guiding proof search. Not only have we been able to rapidly reconstruct a wide range of search-control techniques within a common framework, but we have also been able to exploit the additional power to abstract different occurrences of terms in different ways, according to their history. Thus, rippling has opened the door to a new paradigm for search control.

7
Conclusions

We have come a long way in our investigation of rippling: from the observation of a common pattern in structural induction proofs, to a new paradigm in proof search. Firstly, we noted that this common pattern could be enforced, rather than merely observed, by inserting meta-level annotations into object-level formulas. These annotations – wave-holes and wave-fronts – marked those parts of formulas that were to be preserved and moved, respectively. Ensuring that rewriting respected these annotations enforced additional constraints during proof search: restricting that search to those parts of the search space that made progress towards using the induction hypothesis to prove the induction conclusion.

Secondly, experimental exploration with these annotations suggested a wealth of ways to extend and generalize the original idea beyond simple structural inductions to more complex forms of induction and to many other kinds of proof. Indeed, whenever proving a goal using one or more structurally similar "givens", rippling could help guide the proof through a potential combinatorial explosion towards a successful conclusion with little or no search.

Thirdly, since rippling imposes such strong expectations on the structure of a proof, any failure of rippling can be analyzed to suggest how to patch an initially failed proof attempt. This productive use of failure often suggests proof patches that had previously been thought beyond the ability of automated reasoners: so-called, "eureka" steps. These may include, for instance, the suggestion of a novel induction rule, a new lemma, a generalization of the original conjecture, or a case split.

Fourthly, we have developed a formal theory of rippling, leading to a deeper understanding of how and why it works. In particular, we can prove the termination of any application of rippling. This termination result is remarkable in at least two respects: (a) it applies to an infinite set of rewrite rules, consisting of all rewrite rules that can be annotated as wave-rules, and

(b) this set may include the same equation twice: annotated both left-to-right and right-to-left. The wave annotations prevent the looping that would cause non-termination in conventional rewriting.

Fifthly, we have presented a family of new applications for the insertion of meta-level annotations into object-level formulas. These applications include basic ordered paramodulation and basic superposition; the tracking of focus during inference; and the analogical reuse of proofs of old theorems to guide the proofs of new ones. From these examples, it is clear that we have only scratched the surface of the use of annotations to guide proof search. We antic-ipate that many more such applications remain to be discovered. It is for this reason that we claim to have developed a new paradigm for proof search.

Lastly, all the above ideas have been extensively tested in practical appli-cations. The INKA system has successfully applied rippling to a wide range of industrial-scale verification problems. Rippling has proved its worth by signifi-cantly reducing search, bringing hard problems into the range of a state-of-the-art inductive theorem prover, without excluding the proof being sought from that search space. The INKA system has also been successfully used to thor-oughly test the family of new applications of annotated reasoning described in the last chapter. The *C l*A*M* system has been used to apply rippling to the wide range of inductive and non-inductive problems, as illustrated, for instance, in Chapter 5. It has also been used to explore the productive use of failure using critics that was described in Chapter 3.

This story is not yet finished. Apart from the many further extensions and generalizations to rippling and its theoretical foundations that constantly present themselves, two major challenges remain. One is to develop new appli-cations of annotated reasoning. From the new applications we have presented in the last chapter, the range and potential of this new paradigm seem enor-mous. We urge readers to see if it can be applied to their current search prob-lems. The other is to further explore the productive use of failure, e.g. arising in new applications of annotated reasoning. The starting-point is a declarative specification of the preconditions of the new proof methods. Systematic falsi-fication of each precondition may then suggest an appropriate patch to recover from such a failure. Further analysis of each failure may suggest how to in-stantiate the patch to the current situation. The experience of rippling shows that such analysis can lead to imaginative new proof steps, usually considered beyond automation.

We hope you have enjoyed reading this book and have come to share some of our enthusiasm for rippling and related forms of annotated reasoning. More importantly, we hope you will find these ideas useful in your own work and will join us in the further exploration and development of rippling.

Appendix 1

An annotated calculus and a unification algorithm

In this appendix, we formalize a specific annotation calculus that is able to deal with all the examples that we have presented in Chapter 6. This calculus is based on a first-order proof calculus and a corresponding unification procedure.

We wish to emphasize that the approach presented is an example of a more general technique to combine annotations and logic calculi. Another example is Hutter & Kohlhase (2000), which describes how to incorporate annotations into a calculus based on higher-order logic. Both approaches share the same principal mechanisms to incorporate annotations into calculi.

A1.1 An annotation calculus

The integration of annotations into a calculus is determined by the definition of annotated substitution, which we have only sketched in Section 6.1. As seen in Section 6.2.3, annotated substitutions instantiate both meta-variables and annotation variables. Note that, in contrast to the formalization in Chapter 4, instantiations of meta-variables are independent of instantiations of annotations, since we have separated signatures and variables for annotation terms and object terms.

To cope with these different kinds of variables, an annotated substitution consists of a substitution for meta-variables as well as a family of substitutions for annotation variables. The definition of annotated substitutions determines the possible ways to inherit information during an inference step. To guarantee that an annotated inference step corresponds to a sound inference step in the non-annotated calculus, an annotated substitution has to denote a "standard" substitution if we erase all annotations.

In an annotated calculus (the annotated rewrite system considered here is just an instance of it) the information flow is realized by using annotation

177

variables to annotate rewrite rules. Each annotation occurring in the instance of a variable has to be an instance of the annotation of the variable. In particular, matching the left-hand side of a rule with a subexpression instantiates the annotation variables. The instantiations of the individual annotation variables denote the particular bits of information in the original formula that are transferred to the new formula. Using a term language to represent annotations, we are free to use these bits of information as building blocks for more complex information.

However, there are restrictions on how the information flow can be organized. Consider a rewrite rule in which a meta-variable U occurs on both sides. Let α be its annotation on the left-hand side, and let $t[\alpha]$ be some annotation term of U on the right-hand side. Matching U^α with an annotated term $f^c(a^d, b^e)$, results, in general, in different instantiations of α inside the annotations c, d, and e. The instantiation of α may depend on the position where α occurs in $f^c(a^d, b^e)$. For instance, we can instantiate $U^{t[\alpha]}$ to $f^{t[c]}(a^{t[d]}, b^{t[e]})$. We can also come up with a solution in which the instantiation of α is independent of the position of the symbol occurrence. In this case, the annotated substitution has to identify the annotations c, d, and e. The question arises whether the instantiation of an annotation variable should depend on the variable's context, i.e. are annotation variables instantiated in different ways depending on whether they occur as the annotation of a function symbol or a variable? There is no single answer to this question.

(i) In the reuse example reference of Section 6.2.4, we needed a notion of instantiation that is independent of the symbols to which annotation variables are attached. We used annotation variables to propagate the position of the rule application to all symbol occurrences of the new subterm. This means that an instantiation of these variables is global to all its occurrences. We call such an annotation variable a *rigid* variable.

(ii) In the examples in Sections 6.2.2 and 6.2.3, we used a term U^α as a placeholder for an arbitrary annotated term. If α must always be instantiated in the same way, then U^α could only be instantiated to annotated terms with identical annotations. Thus, there would be no possible annotation of a variable that allows us to instantiate the resulting annotated term to an arbitrary annotated term. Hence, there is also a need for what we will call *flexible* annotation variables that can be instantiated in different ways, depending, for instance, on the symbol to which they are attached. If the annotation variable α occurs in the annotation of a variable U that is instantiated to a compound term, then it is helpful to consider separately the instantiation of α to the annotations of the different symbols in that compound term. For example, when instantiating U^α to $s^{\mathsf{c}}(x^{\mathsf{b}})$ we have to distinguish between the instantiation of α to obtain the annotation c of the top-level symbol s, and the instantiation of α to obtain the annotation b of x.

Our formal definition of annotation variables allows us to handle both kinds of examples.

Definition 16 *The set of* annotation variables $\mathcal{V}$ *consists of two disjoint infinite sets of* rigid *(annotation) variables* $\mathcal{V}_R$ *and* flexible *(annotation) variables* $\mathcal{V}_F$.

Abusing notation, we write $\mathcal{V}_F(t)$ $(\mathcal{V}_R(t))$ to denote all flexible (rigid) variables of an annotated term or an annotation t; $\mathcal{V}_{F,a}(t) \subset \mathcal{V}_F$ denotes all flexible variables occurring in the annotations of symbol a in t, and $\mathcal{V}_a$ is defined by $\mathcal{V}_a(t) := \mathcal{V}_R(t) \cup \mathcal{V}_{F,a}(t)$. We use the expression $\alpha \in \mathcal{V}_a(t)$ to denote that α is either a rigid variable occurring in t, or a flexible variable occurring in some annotation of a in t.

Annotated substitutions are split into an $\mathcal{X}$-substitution operating on meta-variables and $\mathcal{V}$-substitutions operating on annotation variables. Since annotated terms are ordinary terms when their annotations are erased, substitutions on annotated terms denote substitutions on terms if we erase all annotations (cf. Section 6.1).

$\mathcal{X}$-substitutions, representing the underlying substitutions on (non-annotated) terms, are defined in the usual way as functions, which are represented as finite sets of variable/term pairs.

Definition 17 *An $\mathcal{X}$-substitution σ is a function from $\mathcal{X}$ to $\mathcal{T}_{\Sigma(\mathcal{X})}$ such that $\sigma(U) = U$ for all but finitely many $U \in \mathcal{X}$.*

Examples of $\mathcal{X}$-substitutions are $\{x/U, s(y)/V\}$ and $\{x/U\}$.

In a similar way, we define $\mathcal{V}$-substitutions on annotated variables. Additionally, $\mathcal{V}$-substitutions must not instantiate rigid variables by annotations containing flexible variables:

Definition 18 *A $\mathcal{V}$-substitution κ is a function from $\mathcal{V}$ to $\mathcal{T}_{\Pi(\mathcal{V})}$ such that $\kappa(\alpha) = \alpha$ for all but finitely many $\alpha \in \mathcal{V}$, and $\alpha \in \mathcal{V}_R$ implies that $\mathcal{V}_F(\kappa(\alpha)) = \emptyset$.*

An example of a $\mathcal{V}$-substitution is $\{\mathsf{b}/\alpha, \mathsf{c}/\beta\}$, whereas $\{\beta/\alpha\}$ is only a $\mathcal{V}$-substitution if either β is rigid or α is flexible.

Since the instantiation of rigid variables is global to all their occurrences, an annotated substitution provides a $\mathcal{V}$-substitution ζ_R to instantiate these variables.

In contrast to rigid variables, the instantiation of flexible variables depends on the symbols they annotate. For each function symbol f we obtain a separate $\mathcal{V}$-substitution ζ_f that defines the instantiation of the flexible variables occurring in the annotations of f.

An annotation c of a meta-variable X containing flexible variables can be considered simply as a pattern for admissible annotations occurring in later instances $\rho(X)$ of the meta-variable. We use the same pattern to compute the annotations of the individual positions in $\rho(X)$. However, we may use different instances of this pattern in different positions. Thus, the instantiation of a flexible variable depends on the underlying meta-variable and on the position of the designated instantiation of that variable. Hence, annotated substitutions provide families of $\mathcal{V}$-substitutions $\zeta_{U,p}$ that instantiate flexible variables occurring in the annotation of a variable U for a specific position p in the instance $\sigma(U)$.

Definition 19 *A $\mathcal{V}$-annotation family ζ with respect to a $\mathcal{X}$-substitution σ is a family of $\mathcal{V}$-substitutions consisting of*

(i) *a $\mathcal{V}$-substitution ζ_R with $DOM(\zeta_R) \subset \mathcal{V}_R$,*
(ii) *$\mathcal{V}$-substitutions ζ_f with $DOM(\zeta_f) \subset \mathcal{V}_F$ for all $f \in \Sigma$, and*
(iii) *$\mathcal{V}$-substitutions $\zeta_{U,p}$ with $DOM(\zeta_{U,p}) \subset \mathcal{V}_F$ for $U \in \mathcal{X}$ and positions $p \in Pos(\sigma(U))$.*

Using these definitions, we define annotated substitutions as follows.

Definition 20 *An annotated substitution $\rho = (\sigma, \zeta)$ is a pair consisting of a $\mathcal{X}$-substitution σ and a $\mathcal{V}$-annotation family ζ with respect to σ.*

Example Consider the first rewrite step in Section 6.2.4. Suppose that α and δ are rigid variables, and that β and γ are flexible variables. When matching $s^\delta(U^\beta) +^\alpha V^\gamma$ with $s^2(x^3) +^1 s^4(y^5)$ we obtain the following annotated substitution (σ, ζ):

$$\sigma = \{x/U, s(y)/V\},$$
$$\zeta_R = \{1/\alpha, 2/\delta\},$$
$$\zeta_f = \{\} \text{ for all } f \in \Sigma,$$
$$\zeta_{U,\epsilon} = \{3/\beta\},$$
$$\zeta_{V,\epsilon} = \{4/\gamma\}, \text{ and}$$
$$\zeta_{V,\langle 1 \rangle} = \{5/\gamma\}.$$

Definition 21 *An annotated substitution* $\rho = (\sigma, \zeta)$ *defines a* mapping *from* $\mathcal{A} \to \mathcal{A}$ *by:*

(i) $\rho(f^{\mathsf{c}}(t_1, \ldots, t_n)) = f^{\zeta_f(\zeta_R(\mathsf{c}))}(\rho(t_1), \ldots, \rho(t_n))$.

(ii) $\rho(U^{\mathsf{c}}) = t$ *with* $erase(t) = \sigma(U)$ *and* $\mathcal{T}_{\Pi(\mathcal{V})}(t/p) = \zeta_{U,p}(\zeta_R(\mathsf{c}))$ *for all* $p \in Pos(t)$.

Notice that Definition 21 is based on the restriction that flexible variables must not occur in the codomain of instantiations of rigid variables. Thus, the successive application of ζ_R-substitutions and ζ_F-substitutions is identical to the application of the combination of both substitutions.

The annotation calculus is an extension of the underlying calculus, since we *add* annotations to individual symbol occurrences and formulate *additional* cases when defining the unification and matching processes. Given an annotated substitution, its $\mathcal{X}$-substitution is exactly the corresponding substitution in the non-annotated case. Definition 21 trivially guarantees that if an annotated substitution is applicable to an annotated term, then its $\mathcal{X}$-substitution is applicable to the erasure of that term. Hence, the annotated calculus always simulates inference steps of the underlying non-annotated calculus, which guarantees the soundness of the approach.

Based on annotated substitutions, we lift the usual notions of first-order unification theory to the annotated case. An annotated substitution $\rho = (\sigma, \zeta)$ is a *renaming* if and only if its components σ and ζ are renamings. The *composition* $\rho' \circ \rho$ of two annotated substitutions ρ and ρ' is the mapping that maps each annotated term t to $\rho'(\rho(t))$. It is straightforward to prove that $\rho' \circ \rho$ is again an annotated substitution.

Definition 21, which describes the application of annotated substitutions, causes some intrinsic problems in connection with most general unifiers. Suppose α is a flexible variable that occurs in an annotated term $U^{\mathsf{k}(\alpha)}$. Then $\mathsf{k}(\alpha)$ represents a pattern for all annotations occurring in some instantiation of $U^{\mathsf{k}(\alpha)}$. Depending on the position in the instantiation of U, we may instantiate α differently. For instance, $s^{\mathsf{k}(\mathsf{a})}(s^{\mathsf{k}(\mathsf{b})}(0^{\mathsf{k}(\mathsf{c})}))$ is an admissible instance of $U^{\mathsf{k}(\alpha)}$ using an annotated substitution (σ, ζ), with $\sigma = \{s(s(0))/U\}$, $\zeta_{U,\epsilon} = \{\mathsf{a}/\alpha\}$, $\zeta_{U,\langle 1 \rangle} = \{\mathsf{b}/\alpha\}$, and $\zeta_{U,\langle 11 \rangle} = \{\mathsf{c}/\alpha\}$. But if we instantiate U by $s(s(0))$ without specifying any substitution of α explicitly, we obtain $s^{\mathsf{k}(\alpha)}(s^{\mathsf{k}(\alpha)}(0^{\mathsf{k}(\alpha)}))$, which cannot be instantiated to $s^{\mathsf{k}(\mathsf{a})}(s^{\mathsf{k}(\mathsf{b})}(0^{\mathsf{k}(\mathsf{c})}))$ because of the clash of a/α and b/α in ζ_s. Copying the annotations attached to a variable U to function symbols, we must rename the $\mathcal{V}_F$ variables in all positions to obtain the intended instance $s^{\mathsf{k}(\beta)}(s^{\mathsf{k}(\gamma)}(0^{\mathsf{k}(\alpha)}))$. This renaming is not built into the notion of annotated substitutions but has to be done *explicitly*. We must explicitly rename the flexible variables occurring in the annotation of a meta-variable that has to be instantiated by the substitution. Definition 23 computes such an explicit renaming when instantiating a meta-variable. Since we aim at a finite representation of annotated substitutions, we are only able to specify renamings for a finite set of flexible variables. Thus, we always restrict the use of annotated substitutions to the finite set of flexible variables occurring in the annotation of a variable.

To keep track of the flexible variables to be considered during unification, we define contexts.

Definition 22 *A* context Γ *is a finite set of terms* U^α, *with* $U \in \mathcal{X}$ *and* $\alpha \in \mathcal{V}_F$. *The required* context $\Gamma(t)$ *of an annotated term* t *is defined by* $\Gamma(t) = \{U^\alpha \mid U \in \mathcal{X}(t) \text{ and } \alpha \in \mathcal{V}_{F,U}(t)\}$. *The* context Γ_U *of a variable* U *with respect to* Γ *is defined by* $\Gamma_U = \{\alpha \mid U^\alpha \in \Gamma\}$.

The following definition provides an annotated substitution corresponding to a substitution $\{t/U\}$ that takes care of such a renaming for a set of flexible variables $\{\alpha_1, \ldots, \alpha_n\}$.

Definition 23 *Let* $x \in \mathcal{X}$, $t \in \mathcal{T}_{\Sigma(\mathcal{X})}$. *An annotated substitution* $(\{t/U\}, \zeta)$ *is an* annotation renaming *of* $\{t/U\}$ *with respect to a context* Γ_U *iff*

  (i) $\zeta_R = \{\}$ *and* $\zeta_f = \{\}$ *for all* $f \in \Sigma$, *and*
 (ii) *for all positions* p, p' *of* t *and* $\alpha \in \Gamma_U$:
    (a) $\zeta_{U,p}(\alpha) \in \mathcal{V}_F \setminus \Gamma_U$, *and*
    (b) $\zeta_{U,p}(\alpha) = \zeta_{U',p'}(\alpha')$ *implies that* $U = U'$, $p = p'$, *and* $\alpha = \alpha'$.

For example, let $\{\alpha, \beta, \gamma\}$ be flexible variables. Then an annotation renaming of $\{s(y)/V\}$ with respect to $\{\alpha\}$ is a pair $(\{s(y)/V\}, \zeta)$ with

$$\zeta_R = \{\}, \zeta_f = \{\} \text{ for all } f \in \Sigma, \text{ and } \zeta_{V,\epsilon} = \{\beta/\alpha\}, \zeta_{V,\langle 1 \rangle} = \{\gamma/\alpha\}.$$

Applying this annotation renaming to an annotated term V^α results in an annotated term $s^\beta(y^\gamma)$, which is annotated with flexible variables that are not in $\Gamma_V = \{\alpha\}$.

As a consequence of the necessary explicit renaming of flexible variables, we must do some bookkeeping about the flexible variables that are in the scope of an explicit renaming. Therefore we define:

Definition 24 *Two annotated substitutions* ρ *and* ρ' *are* equal *with respect to a context* Γ, *written* $\rho =_\Gamma \rho'$ *for short, iff* $\rho(t) = \rho'(t)$ *for all* $t \in \mathcal{A}$ *with* $\Gamma(t) \subset \Gamma$.

We are now ready to introduce the notion of most general annotated unifier:

Definition 25 *An annotated substitution ρ is an* annotated unifier *of two annotated terms s and t iff $\rho(s) = \rho(t)$; ρ is a most-general annotated unifier of s and t with respect to a context Γ iff for all unifiers ρ' of s and t there exists an annotated substitution λ with $\lambda \circ \rho =_{\Gamma} \rho'$.*

A1.2 Unification algorithm

In the following we develop a unification algorithm for first-order annotated terms. As usual, we define the unification algorithm as a set of transformation rules operating on a set of annotated term pairs denoting the unification problem. Each member of this set is either a pair of annotated terms or a pair $a^{\mathsf{c}} = a^{\mathsf{d}}$ of annotated function or variable symbols. The latter kind represents a unification problem that is purely concerned with annotation variables. The symbol a is used to denote the context in which the annotated unification problem occurs. Unifying this pair imposes constraints on the corresponding substitutions of flexible variables ζ_a.

Definition 26 *An annotated substitution ρ is an* annotated unifier *of a unification problem $E = \{s_1 = t_1, \ldots, s_n = t_n\}$ iff $\rho(s_i) = \rho(t_i)^1$ for all $1 \leq i \leq n$. The set of annotated unifiers of E is denoted by $\mathcal{U}(E)$.*

A unification problem is simplified by iteratively applying transformation rules to the problem until the result is in a so-called solved form, i.e. we can easily read the resulting unifier from this set or a clash occurs, i.e. there is a pair that is not unifiable.

A solved form consists of two different kinds of equations. The first kind denotes the usual $\mathcal{X}$-substitutions, i.e. a variable U is replaced by a term t. Additionally, we encode the $\mathcal{V}$-substitutions on flexible variables occurring in the annotations of U into such equations. Thus, for each annotation variable α under consideration, we obtain an equation $U^{\alpha} = t'$ with $erase(t') = t$. The annotations in t' specify the different instantiations of α with respect to the positions in t'. The second kind of equation is concerned with the instantiation of rigid annotation variables and flexible annotation variables attached to function symbols. We denote such instantiations by equations like $a^{\alpha} = a^{\mathsf{d}}$.

Definition 27 *A pair L is in a* solved form *with respect to a set of equations E and a context Γ iff*

---

^{1} We extend the application of annotated substitutions to term fragments a^{c} in the obvious way.

(i) *L has the form* $U^\alpha = t$ *with* $U \in \mathcal{X} \setminus \mathcal{X}(t)$ *and* $\alpha \in \Gamma_U$. *Furthermore, for all* $\beta \in \Gamma_U$ *there is exactly one literal* $U^\beta = t' \in E$ *with* $erase(t) = erase(t')$ *and* U *occurs nowhere else in* E.

(ii) *L has the form* $a^\alpha = a^c$, $a \in \Sigma \cup \mathcal{X}$, $\alpha \in \mathcal{V}$, *and* $\alpha \notin \mathcal{V}_a(c) \cup \mathcal{V}_a(E)$.

The definition of solved forms reflects the different kinds of variables. The first clause deals with term variables. All pairs containing U must share a common erasure, and U must not occur in t. In general, we need different pairs for U to encode the instantiation of the flexible variable attached to U. The second clause deals with annotation variables. Since the instantiation of flexible variables depends on the symbol they are attached to, the unification algorithm operates on annotated symbols. To simplify matters, we use the same representation for rigid and flexible variables. The condition of the second clause guarantees that we do not obtain more than one pair for the same rigid variable.

Definition 28 *A unification problem* E *is in* solved form *with respect to a context* Γ *iff all its pairs* L *are in solved form with respect to* $E \setminus \{L\}$ *and* Γ.

For example, the unification problem

$$\{s^\delta = s^2, +^\alpha = +^1, U^\beta = x^3, V^\beta = s^{\beta'}(y^{\beta''}), V^\gamma = s^4(y^5)\} \qquad (\text{A1.1})$$

is in solved form with respect to $\{U^\beta, V^\beta, V^\gamma\}$. Notice that this unification problem is not solved with respect to the context $\{U^\beta, V^\beta, V^\gamma, V^\delta\}$ since it does not provide an instantiation for δ attached to V. A unification problem in solved form denotes an annotated substitution.

Definition 29 *Given a unification problem* E *in solved form with respect to a context* Γ, *then the* canonical annotated substitution $\rho_E = (\sigma, \zeta)$ *of* E *with respect to* Γ *is defined as follows:*

(i) $\sigma(U) = erase(t)$ *if some literal* $U^\alpha = t \in E$, *otherwise* $\sigma(U) = U$.

(ii) $\zeta_R(\alpha) = c$ *if* $a^\alpha = a^c \in E$ *with* $a \in \Sigma \cup \mathcal{X}$ *and* $\alpha \in \mathcal{V}_R$; *otherwise* $\zeta_R(\alpha) = \alpha$.

(iii) $\zeta_f(\alpha) = c$ *if* $f^\alpha = f^c \in E$ *with* $\alpha \in \mathcal{V}_F$ *and* $f \in \Sigma$, *otherwise* $\zeta_f(\alpha) = \alpha$.

(iv) $\zeta_{U,p}(\alpha) = c$ *if* $U^\alpha = t \in E$ *with* $\alpha \in \mathcal{V}_F$ *and* $\mathcal{T}_{\Pi(V)}(t/p) = c$; *otherwise* $\zeta_{U,p}(\alpha) = \alpha$.

The unification problem (A1.1) provides the following canonical annotated substitution (σ, ζ):

(i) $\sigma = \{x/U, s(y)/V\}$.

(ii) $\zeta_R = \{1/\alpha, 2/\delta\}$.

(iii) $\zeta_f = \{\}$ for all $f \in \Sigma$.

(iv) $\zeta_{U,\epsilon} = \{3/\beta\}$.

(v) $\zeta_{V,\epsilon} = \{\beta'/\beta, 4/\gamma\}$, and $\zeta_{V,\langle 1\rangle} = \{\beta''/\beta, 5/\gamma\}$.

Obviously, the canonical annotated substitution ρ_E with respect to Γ is a most general annotated unifier of E with respect to Γ. The definition of a solved form ensures that the definition of the canonical annotated substitution is unique up to variable renaming.

We use this definition to denote annotated substitutions as a set of replacements $\{t_1/s_1, \ldots, t_n/s_n\}$ iff the corresponding unification problem $\{s_1 = t_1, \ldots, s_n = t_n\}$ is in solved form.

We are now ready to introduce the transformation rules of the unification procedure. Each rule will operate on a pair E ; Γ consisting of a unification problem E and a context Γ. We apply these rules with the understanding that the operator $=$ is symmetric and that trivial pairs may be dropped. Finally, no rule may be applied to a solved pair.

The first two transformation rules decompose a unification problem based on the structure of the terms or their annotations. Since both $\mathcal{X}$-substitutions and $\mathcal{V}$-substitutions are homomorphic extensions of mappings on $\mathcal{X}$ and $\mathcal{V}$, the unification of compound terms or annotations can be reduced to the unification of their components:

Σ-*Decomposition*:

$$\frac{\{f^{\mathsf{c}}(t_1, \ldots, t_n) = f^{\mathsf{d}}(s_1, \ldots, s_n)\} \cup E \; ; \; \Gamma}{\{f^{\mathsf{c}} = f^{\mathsf{d}}, t_1 = s_1, \ldots, t_n = s_n\} \cup E \; ; \; \Gamma}$$

Π-*Decomposition*:

$$\frac{\{a^{\mathsf{k}(t_1, \ldots, t_n)} = a^{\mathsf{k}(s_1, \ldots, s_n)}\} \cup E \; ; \; \Gamma}{\{a^{t_1} = a^{s_1}, \ldots, a^{t_n} = a^{s_n}\} \cup E \; ; \; \Gamma}$$

with $a \in \Sigma \cup \mathcal{X}$ and $\mathsf{c}, \mathsf{d} \in \mathcal{T}_{\Pi(\mathcal{V})}$.

Both rules allow us to decompose unification problems until we either encounter a clash, i.e. there is an equation with two different function symbols on left-hand and right-hand sides, or with a pair where at least one side consists of a term variable or an annotation variable. Owing to the homomorphic extension of $\mathcal{X}$-substitutions and $\mathcal{V}$-substitutions to annotated terms, it is easy to see that both transformations will keep the set of unifiers unchanged.

For example, consider the previous unification problem

$$\{s^\delta(U^\beta) +^\alpha V^\gamma = s^2(x^3) +^1 s^4(y^5)\}; \{U^\beta, V^\beta, V^\gamma\}. \qquad \text{(A1.2)}$$

Applying the decomposition rules results in the unification problem

$$\{s^\delta = s^2, +^\alpha = +^1, U^\beta = x^3, V^\gamma = s^4(y^5)\}; \{U^\beta, V^\beta, V^\gamma\}. \qquad \text{(A1.3)}$$

Annotation variables are treated in the unification process like first-order term variables in standard unification theory. Suppose there is an annotation equation $a^\alpha = a^c$ in the problem set E. Either α occurs in c and thus both terms are not unifiable, or we replace each relevant (depending whether α is rigid or flexible) occurrence of α in E by c. As an additional restriction we must guarantee that if α is a rigid variable then c does not contain any flexible variables.

$\mathcal{V}$-*Substitution*:

$$\frac{\{a^\alpha = a^c\} \cup E \; ; \; \Gamma}{\{a^\alpha = a^c\} \cup \{a^c/a^\alpha\}(E) \; ; \; \Gamma}$$

if $a \in \Sigma \cup \mathcal{X}, \alpha \in \mathcal{V}_a(E) \setminus \mathcal{V}(c)$ and $\alpha \in \mathcal{V}_R$ implies $\mathcal{V}_F(c) = \emptyset$.

In the example of (A1.3), the $\mathcal{V}$-Substitution rule is not applicable since both δ and α only occur once in already solved pairs.

If α is a rigid variable and c contains flexible variables, then the above rule is not applicable because the rigid variables may not be instantiated by flexible annotations. But we can, of course, instantiate the flexible variables of c to rigid variables. Thus, we imitate the annotation of the right-hand side until we succeed in isolating the flexible variables occurring on the right-hand side.

Π-*Imitation*:

$$\frac{\{a^\alpha = a^{k(t_1,\ldots,t_n)}\} \cup E \; ; \; \Gamma}{\{a^\alpha = a^{k(\alpha_1,\ldots,\alpha_n)}, a^{k(\alpha_1,\ldots,\alpha_n)} = a^{k(t_1,\ldots,t_n)}\} \cup E \; ; \; \Gamma}$$

if $a \in \Sigma \cup \mathcal{X}, \alpha \in \mathcal{V}_R, \mathcal{V}_F(k(t_1,\ldots,t_n)) \neq \emptyset$, and $\alpha_1,\ldots,\alpha_n$ are fresh, pairwise distinct rigid annotation variables.

We are left with the case of $U^c = t$. Obviously, this equation requires that the term substitution σ unifies U and $erase(t)$. Analogous to standard first-order unification, we will simplify the problem set by replacing occurrences of U by appropriate annotated versions of $erase(t)$. The problem remains of how to annotate $erase(t)$ to replace an occurrence U^c. If c does not contain any flexible variables, then the definition of annotated substitutions requires that each symbol of t is annotated with the same instance of c. If c contains a flexible variable α, then we are free to instantiate α differently for each position in $erase(t)$. Thus, to obtain a most general unifier, we have to rename α in c by new *fresh* flexible variables for each position in $erase(t)$. We use the definition of annotated renamings to replace all (annotated) occurrences of a term variable U in the problem set, once we find an equation $U^c = t$.

$\mathcal{X}$-*Substitution*:

$$\frac{\{U^c = t\} \cup E \; ; \; \Gamma}{\{U^\alpha = \tau(U^\alpha) | \alpha \in \Gamma_U\} \cup \{\tau(U^c) = \tau(t)\} \cup \tau(E) \; ; \; \Gamma \cup \bigcup_{\alpha \in \Gamma_U} \Gamma(\tau(U^\alpha))}$$

if $U \in \mathcal{X} \setminus \mathcal{X}(t)$, $U^c = t$ is not in solved form with respect to E and τ is an annotation renaming of $\{erase(t)/U\}$ with respect to Γ_U.

The pairs $\{U^\alpha = \tau(U^\alpha) | \alpha \in \Gamma_U\}$ represent the instantiation of the variants of U by variants of t that determine the substitutions $\zeta_{U,p}$, for all the positions p in t. Since the annotated renaming τ introduces new flexible variables, we have to extend the context Γ by the newly introduced annotated variables.

Consider our example of solving the unification problem (A1.3). If we select the equation $U^\beta = x^3$ to apply the $\mathcal{X}$-Substitution rule, then we obtain an annotated renaming τ with $\tau_{V,\epsilon} = \{\beta'/\beta, \gamma'/\gamma\}$ and $\tau_{V,\langle 1 \rangle} = \{\beta''/\beta, \gamma''/\gamma\}$. Since $\tau(U^\alpha)$ does not contain any term variables, $\Gamma(\tau(U^\alpha))$ is empty. Thus, we obtain

$$\{V^\gamma = s^{\gamma'}(y^{\gamma''}), V^\beta = s^{\beta'}(y^{\beta''}), s^\delta = s^2,$$
$$+^\alpha = +^1, U^\beta = x^3, s^{\gamma'}(y^{\gamma''}) = s^4(y^5)\}; \{U^\beta, V^\beta, V^\gamma\}. \quad \text{(A1.4)}$$

Applying the Σ-Decomposition rule yields

$$\{V^\gamma = s^{\gamma'}(y^{\gamma''}), V^\beta = s^{\beta'}(y^{\beta''}), s^\delta = s^2,$$
$$+^\alpha = +^1, U^\beta = x^3, s^{\gamma'} = s^4, y^{\gamma''} = y^5\}; \{U^\beta, V^\beta, V^\gamma\}. \quad \text{(A1.5)}$$

After applying the $\mathcal{V}$-Substitution rule twice we finally obtain the solved form

$$\{V^\gamma = s^4(y^5), V^\beta = s^{\beta'}(y^{\beta''}), s^\delta = s^2,$$
$$+^\alpha = +^1, U^\beta = x^3, s^{\gamma'} = s^4, y^{\gamma''} = y^5\}; \{U^\beta, V^\beta, V^\gamma\}. \quad \text{(A1.6)}$$

All the presented transformation rules form the unification algorithm:

Definition 30 *The rules $\mathcal{R}$ for annotated unification are Σ-Decomposition, Π-Decomposition, $\mathcal{V}$-Substitution, Π-Imitation, and $\mathcal{X}$-Substitution.*

Given two sets E, E' of equations and two contexts Γ, Γ' then $E; \Gamma \vdash E'; \Gamma'$ holds iff

(i) *$E = E'$ and $\Gamma = \Gamma'$; or*
(ii) *there is a set of equation E'' and a context Γ'' such that $\frac{E;\Gamma}{E'';\Gamma''} \in \mathcal{R}$ and $E''; \Gamma'' \vdash E'; \Gamma'$ holds.*

A1.2.1 Soundness and completeness

We now prove the soundness and completeness of the annotated unification algorithm in the usual way. We prove that the set of annotated unifiers for a unification problem is invariant with respect to the transformation rules.

As a prerequisite, we prove a lemma that guarantees that applying explicit renamings of flexible variables does not change a unification problem in an essential way. In particular, suppose σ is an annotated unifier of U and t, and ρ is an annotated renaming of $\{t/U\}$ with respect to some context Γ. Then our lemma states that we can reduce the unification problem $U = t$ to a unification problem $\rho(U) = t$ without losing any potential unifiers.

Lemma 6 *Let $U \in \mathcal{X}$, $t \in \mathcal{A}$, and $\tau = (\lambda, \kappa)$ be an annotation renaming of $\{erase(t)/U\}$ with respect to some context Γ. Let $\rho = (\sigma, \zeta)$ be an annotated substitution with $\sigma(U) = \sigma(t)$. Then there is an annotated substitution $\rho' = (\sigma', \zeta')$ with $\rho =_\Gamma \rho' \circ \tau$.*

Proof We sketch the proof by defining an appropriate annotated substitution $\rho' = (\sigma', \zeta')$ satisfying the requirements of lemma 6.

Let $\sigma' = \sigma$. Then, obviously, $\sigma = \sigma' \circ \lambda$ holds. Let ζ' be ζ except for those flexible variables that occur in the codomain of κ. Let β be such a variable. There is a unique annotation variable $\alpha \in \mathcal{V}_F$ and position $p \in Pos(\lambda(U))$ such that $\mathcal{T}_{\Pi(\mathcal{V})}(\tau(U^\alpha)/p) = \beta$ holds. If $\lambda(U)/p$ is a variable V, then we (re-)define $\zeta'_{V,p'}(\beta) = \zeta_{U,p'\circ p}(\alpha)$ for all $p' \in Pos(\sigma(V))$. If $\lambda(U)/p$ is a function symbol $f \in \Sigma$ then we define $\zeta'_f(\beta) = \zeta_{U,p}(\alpha)$. □

We are now ready to prove the soundness of the annotated unification algorithm. The usual proviso is that the unification problem does not contain variables that are introduced by one of the transformation rules as a "new" variable. Hence we postulate:

Theorem 7 *Let E; $\Gamma \vdash_\mathcal{R} E'$, Γ'. Then $\mathcal{U}(E) =_\Gamma \mathcal{U}(E')$.*

Proof sketch: We prove this theorem by induction on the length of the transformation. Suppose there is a transformation $E = E_1 \to \ldots \to E_n = E'$.

In the base case, for $n = 1$, the postulated theorem holds trivially. In the induction step we assume that the property holds for all steps except the last one, and perform a case analysis according to the last transformation rule used.

Obviously, the Σ-Decomposition and Π-Decomposition rules keep the set of unifiers invariant because of the definition of annotated substitutions as homomorphic extensions on mappings on term and annotation variables.

In the case of the $\mathcal{V}$-Substitution rule, we employ the standard proof technique for variable elimination from first-order unification theory. Each unifier of $a^\alpha = a^c$ has to identify α and c either globally, if $\alpha \in \mathcal{V}_R$, or locally to occurrences of a, if $\alpha \in \mathcal{V}_F$.[1] Thus, we can replace occurrences of α by c without changing the annotated unifier set, since $a^\alpha = a^c$, which is also a member of the new unification problem, enforces the identification of both.

Consider next the $\mathcal{X}$-Substitution rule, which replaces all occurrences of U by annotated versions of t using an annotated renaming τ. Suppose that there is an annotated unifier ρ of E_{n-1}. Then, according to lemma 6, there is also an annotated substitution ρ' such that $\rho =_\Gamma \rho' \circ \tau$, and ρ' unifies also each $U^{\alpha_i} = \tau(U^\alpha)$ with $\alpha \in \Gamma_U$. Hence, $\rho' \circ \tau$ is a unifier of E_n. Suppose there is an annotated unifier ρ of E_n, then obviously ρ is also an annotated unifier of E_{n-1} □

As mentioned above, the $\mathcal{X}$-Substitution rule must keep track of the necessary renamings of flexible variables occurring in the annotations of the variable to be replaced.

Theorem 7 guarantees that applying the transformation rules will keep the set of unifiers invariant. In order to assure completeness of the approach we must also prove that the unification procedure terminates. For each unification problem, the algorithm will terminate with a unification problem that is either in solved form or contains a clash. This proof is analogous to corresponding proofs in standard unification theory using lexicographical ordering combining the ordering on the number of variables that occur in non-solved forms and the complexity of the occurring terms.

To ease readability, we have presented a simple version of the unification algorithm and have not incorporated any improvements, for instance more sophisticated book keeping of flexible variables.

---

[1] We do not need to distinguish both cases in the rule because the notation of $\mathcal{V}_a$ takes care of this.

Appendix 2

Definitions of functions used in this book

- $X : nat + Y : nat$ computes the sum of two naturals:

$$X + Y = Y$$
$$s(X) + Y = s(X + Y)$$

- $X : nat \times Y : nat$ computes the product of two naturals:

$$0 \times Y = 0$$
$$s(X) \times Y = Y + (X \times Y)$$

- $half(X : nat)$ computes the half of a natural:

$$half(0) = 0$$
$$half(s(0)) = 0$$
$$half(s(s(X))) = s(half(X))$$

- $odd(X : nat)$ checks whether a natural X is odd:

$$odd(0) \leftrightarrow false$$
$$odd(s(0)) \leftrightarrow true$$
$$odd(s(s(X))) \leftrightarrow odd(X)$$

- $even(X : nat)$ checks whether a natural X is even:

$$even(0) \leftrightarrow true$$
$$even(s(0)) \leftrightarrow false$$
$$even(s(s(X))) \leftrightarrow even(X)$$

- *binom(X : nat, Y : nat)* computes the binomial:

$$binom(X, 0) = s(0)$$
$$binom(0, s(Y)) = 0$$
$$binom(s(X), s(Y)) = binom(X, s(Y)) + binom(X, Y)$$

- *length(X : list)* computes the length of a list X:

$$length(nil) = 0$$
$$length(X :: Y) = s(length(Y))$$

- $X : list <> Y : list$ concatenates the two lists X and Y:

$$nil <> Z = Z$$
$$(X :: Y) <> Z = X :: (Y <> Z)$$

- *del(X : elem, Y : list)* deletes the first occurrence of X in Y:

$$del(X, nil) = nil$$
$$X = Y \rightarrow del(X, Y :: Z) = Z$$
$$X \neq Y \rightarrow del(X, Y :: Z) = Y :: del(X, Z)$$

- *rotate(X : nat, Y : list)* moves the first X elements of Y to the end of Y:

$$rotate(0, Z) = Z$$
$$rotate(s(X), nil) = nil$$
$$rotate(s(X), Y :: Z) = rotate(X, (Z <> (Y :: nil)))$$

- *rev(X : list)* reverses the list X

$$rev(nil) = nil$$
$$rev(X :: Y) = rev(Y) <> (X :: nil)$$

- *qrev(Y : list, Z : list)* appends the reversed list of Y in front of Z:

$$qrev(nil, Z) = Z$$
$$qrev(X :: Y, Z) = qrev(Y, X :: Z)$$

- *permute(Y : list, Z : list)* tests whether Y is a permutation of Z

$$permute(nil, Z) \leftrightarrow Z = nil$$
$$X \notin Z \rightarrow permute(X :: Y, Z) \leftrightarrow false$$
$$X \in Z \rightarrow permute(X :: Y, Z) \leftrightarrow permute(Y, del(X, Z))$$

- *palin*($X : list, Y : list$) appends X and the reverse of X to Y:

$$palin(nil, Acc) = Acc$$
$$palin(H :: T, Acc) = H :: palin(T, H :: Acc)$$

- *size*($X : tree$) computes the number of leafs in a tree:

$$size(leaf) = s(0)$$
$$size(node(N, L, R)) = s(size(L) + size(R))$$

- *nodes_in* flattens a tree to a multi-set:

$$nodes\_in(empty) = nil$$
$$nodes\_in(node(N, L, R)) = insert(N, nodes\_in(L) \cup nodes\_in(R))$$

- *count* computes the number of elements in a multi-set:

$$count(empty) = 0$$
$$count(insert(E, S)) = s(count(S))$$

References

Armando, A., Gallagher, J., Smaill, A. and Bundy, A. (1998). Automating the synthesis of decision procedures in a constructive metatheory. *Annals of Mathematics and Artificial Intelligence*, 22(3–4), 259–79. Also available as Research Paper no. 934, DAI, University of Edinburgh.

Armando, A., Smaill, A. and Green, I. (1999). Automatic synthesis of recursive programs: the proof-planning paradigm. *Automated Software Engineering*, pp. 329–56.

Aubin, R. (1975). Some generalization heuristics in proofs by induction. In G. Huet & G. Kahn, eds., *Actes du Colloque Construction: Amélioration et vérification de Programmes*. Institut de recherche d'informatique et d'automatique.

Aubin, R. (1976). *Mechanizing Structural Induction*. Unpublished Ph.D. thesis, University of Edinburgh.

Autexier, S. and Hutter, D. (1997). Equational proof-planning by dynamic abstraction. In U. Furbach & M. P. Bonacina, eds., *International Workshop on First-Order Theorem Proving – FTP97*, pp. 1–6, Linz. RISC-Linz Report Series No. 97–50.

Autexier, S. and Hutter, D. (2002). Maintenance of formal software developments by stratified verification. In *Proceedings 9th International Conference on Logic for Programming Artificial Intelligence and Reasoning*. Springer-Verlag, LNAI.

Autexier, S. (2003). *Hierarchical Contextual Reasoning*. Unpublished Ph.D. thesis, Saarland University / DFKI.

Autexier, S., Hutter, D., Mantel, H. and Schairer, A. (1999). System description: INKA 5.0 – a logical voyager. In H. Ganzinger, ed., *Proceedings 16th International Conference on Automated Deduction, CADE-16*, Trento. Springer-Verlag, LNAI 1632.

Autexier, S., Hutter, D., Langenstein, B., Mantel, H., Rock, G., Schairer, A., Stephan, W., Vogt, R. and Wolpers, A. (2000). VSE: Formal methods meet industrial needs. *International Journal on Software Tools for Technology Transfer, Special Issue on Mechanized Theorem Proving for Technology*, 3(1), 66–77.

Autexier, S., Hutter, D., Mossakowski, T. and Schairer, A. (2002). The development graph manager MAYA. In *Proceedings 9th International Conference on Algebraic Methodology And Software Technology, AMAST2002*. Springer-Verlag, LNCS 2422.

Baader, F. and Nipkow, T. (1998). *Term Rewriting and all That*. Cambridge University Press, Cambridge.

Bachmair, L., Ganzinger, H., Lynch, C. and Snyder, W. (1992). Basic paramodulation and superposition. In D. Kapur, ed., *Proceedings of the 11th International Conference on Automated Deduction (CADE–11)*, pp. 462–76, Saratoga Springs, NY. Springer-Verlag, LNAI 607.

Barnes, J. (2003). High Integrity Software: The SPARK Approach to Safety and Security. Addison-Wesley.

Basin, D. A. and Walsh, T. (1993). Difference unification. In R. Bajcsy, ed., *Proceedings 13th International Joint Conference on Artificial Intelligence (IJCAI '93)*, vol. 1, pp. 116–22, ed. R. Bajcsy. San Mateo, CA. Morgan Kaufmann. Also available as Technical Report MPI-I-92-247, Max-Planck-Institut für Informatik.

Basin, D. and Walsh, T. (1996). A calculus for and termination of rippling. *Journal of Automated Reasoning*, 16(1–2), 147–80.

BBC. (1996). *Fermat's Last Theorem*. Broadcasting Support Services for BBC Horizon, British Broadcasting Corporation: Transcript of a BBC Horizon programme first shown on 15 January 1996.

Benzmüller et al., C. (1997). Ωmega: towards a methematical assistant. In W. McCune, ed., *14th International Conference on Automated Deduction*, pp. 252–9. Springer-Verlag. Also available as SEKI-Report SR-95-05.

Bläsius, K. H. and Siekmann, J. H. (1988). Partial unification for graph-based equational reasoning. In R. Lusk and R. Overbeek eds., *9th International Conference on Automated Deduction*, pp. 397–414. Springer-Verlag.

Bledsoe, W. W. (1990). Challenge problems in elementary calculus. *Journal of Automated Reasoning*, 6(3), 341–59.

Bledsoe, W. W., Boyer, R. S. and Henneman, W. H. (1972). Computer proofs of limit theorems. *Artificial Intelligence*, 3, 27–60.

Boyer, R. S. and Moore, J. S. (1979). *A Computational Logic*. Academic Press, ACM monograph series.

Bundy, A. (1988). The use of explicit plans to guide inductive proofs. In R. Lusk, and R. Overbeek, eds., *9th International Conference on Automated Deduction*. Springer-Verlag, pp. 111–20. Longer version available from Edinburgh as DAI Research Paper No. 349.

Bundy, A. (1991). A science of reasoning. In J.-L. Lassez, and G. Plotkin, eds., *Computational Logic: Essays in Honor of Alan Robinson*. MIT Press, pp. 178–98. Also available from Edinburgh as DAI Research Paper 445.

Bundy, A. (2002a). The termination of rippling + unblocking. University of Edinburgh, Informatics research paper.

Bundy, A. (2002b). Towards the integration of difference removal and difference moving. University of Edinburgh, Informatics technical report.

Bundy, A., van Harmelen, F., Hesketh, J., Smaill, A. and Stevens, A. (1989). A rational reconstruction and extension of recursion analysis. In N. S. Sridharan, ed., *Proceedings of the Eleventh International Joint Conference on Artificial Intelligence*. Morgan Kaufmann, pp. 359–65. Also available from Edinburgh as DAI Research Paper 419.

Bundy, A. and Lombart, V. (1995). Relational rippling: a general approach. In *Proceedings of IJCAI-95*, ed. C. Mellish, pp. 175–81. IJCAI.

Bundy, A., Smaill, A. and Hesketh, J. (1990a). Turning eureka steps into calculations in automatic program synthesis. In S. L. H. Clarke, ed., *Proceedings of UK IT 90*. IEE, pp. 221–6. Also available from Edinburgh as DAI Research Paper 448.

Bundy, A., van Harmelen, F., Horn, C. and Smaill, A. (1990b). The Oyster-Clam system. In M. E. Stickel, ed., *10th International Conference on Automated Deduction*. Springer-Verlag, pp. 647–8. Lecture Notes in Artificial Intelligence No. 449. Also available from Edinburgh as DAI Research Paper 507.

Bundy, A., van Harmelen, F., Smaill, A. and Ireland, A. (1990c). Extensions to the rippling-out tactic for guiding inductive proofs. In M. E. Stickel, ed., *10th International Conference on Automated Deduction*. Springer-Verlag, pp. 132–46. Lecture Notes in Artificial Intelligence No. 449. Also available from Edinburgh as DAI Research Paper 459.

Bundy, A., Stevens, A., van Harmelen, F., Ireland, A. and Smaill, A. (1993). Rippling: A heuristic for guiding inductive proofs. *Artificial Intelligence*, **62**, 185–253. Also available from Edinburgh as DAI Research Paper No. 567.

Cleve, J. and Hutter, D. (1994). A methodology for equational reasoning. In Jay F. Nunamaker, Jr. and Ralph H. Sprague, Jr., (eds.), *Proceedings Hawaii International Conference on System Sciences 27*, vol. III: *Information Systems: DSS/Knowledge-based Systems*. IEEE, Los Alamitos, Ca: Computer Society Press, pp. 569–78.

Cresswell, S., Smaill, A. and Richardson, J. D. C. (1999). Deductive synthesis of recursive plans in linear logic. In *Proceedings of the 5th European Conference on Planning, Durham, UK*, vol. 1809 of *LNAI*. Springer-Verlag.

Dershowitz, N. (1987). Termination of rewriting. In J.-P., Jouannaud, ed., *Rewriting Techniques and Applications*. Academic Press, pp. 69–116.

Digricoli, V. J. (1979). Resolution by unification and equality. In W. H., Joyner, ed., *Proceedings of the Fourth Workshop on Automated Deduction*, pp. 43–52.

Digricoli, V. J. (1980). First experiments with rue automated deduction. In R. Balzer, ed., *Proceedings of the 1st Annual National Conference on Artificial Intelligence*. Stanford University: William Kaufmann, pp. 96–8.

Digricoli, V. J. (1994). The RUE theorem-proving system: the complete set of Lim+ challenge problems. *Journal of Automated Reasoning*, **12**(2), 241–64.

Dixon, L. and Fleuriot, J. D. (2003). IsaPlanner: A Prototype Proof Planner in Isabelle. In F. Baader, ed., *19th International Conference on Automated Deduction*. Springer-Verlag, pp. 279–283. Lecture Notes in Artificial Intelligence No. 2741.

Feigenbaum, E. A. and Feldman, F. (1963). *Computers and Thought*. McGraw-Hill.

Floyd, R. W. (1967). Assigning meanings to programs. In J. T. Schwartz, (ed.), *Mathematical Aspects of Computer Science, Proceedings of Symposia in Applied Mathematics 19*. American Mathematical Society, pp. 19–32.

Gelernter, H. (1963). Realization of a geometry theorem-proving machine. In E. Feigenbaum, and J. Feldman, eds., *Computers and Thought*. McGraw Hill, pp. 134–52.

Gentzen, G. (1934). Untersuchen über das logische Schließen. *Mathematische Zeitschrift*, **39**, 179–210, 405–31.

Giunchiglia, F. and Walsh, T. (1992). A theory of abstraction. *Artificial Intelligence*, **57**(2–3), 323–89.

Gow, J. (2004). *The Dynamic Creation of Induction Rules Using Proof Planning*. School of Informatics, University of Edinburgh.

Gries, D. (1981). *The Science of Programming*. New York, Springer-Verlag.

Guard, J., Oglesby, F., Bennett, J. and Settle, L. (1969). Semi-automated mathematics. *Association for Computing Machinery*, **16**, 49–62.

Harper, R., Honsell, F. and Plotkin, G. (1992). A framework for defining logics. *Journal of the ACM*, **40**(1), 143–84. Preliminary version in LICS 1987.

Hesketh, J. T. (1991). *Using Middle-Out Reasoning to Guide Inductive Theorem Proving*. Unpublished Ph.D. thesis, University of Edinburgh.

Hoare, C. A. R. (1969). An axiomatic basis for computer programming. *Communications of the ACM*, **12**, 576–83.

Hobbs, J. R. (1985). Granularit. In A. Joshi, ed., *Proceedings of the Ninth International Joint Conference on Artificial Intelligence*, Los Altos: Morgan Kaufmann.

Hummel, B. (June 1987). An investigation of formula generalisation heuristics for induction proofs. Interner Bericht 6/87, Universitaet Karlsruhe.

Hutter, D. (1990). Guiding inductive proofs. In M. E. Stickel, ed., *10th International Conference on Automated Deduction*. Springer-Verlag, pp. 147–61. Lecture Notes in Artificial Intelligence No. 449.

Hutter, D. (1997). Colouring terms to control equational reasoning. *Journal of Automated Reasoning*, **18**, 399–442.

Hutter, D. and Kohlhase, M. (1997). A colored version of the λ-Calculus. In McCune, W., ed., *14th International Conference on Automated Deduction*. Springer-Verlag, pp. 291–305. Also available as SEKI-Report SR-95-05.

Hutter, D. and Kohlhase, M. (2000). Managing structural information by higher-order colored unification. *Journal of Automated Reasoning*, **25**(2), 123–64.

Hutter, D. and Sengler, C. (1996). INKA: the next generation. In M. A. McRobbie, and J. K. Slaney, eds., *13th International Conference on Automated Deduction*. Springer-Verlag, pp. 288–92. Springer Lecture Notes in Artificial Intelligence No. 1104.

Ireland, A. and Bundy, A. (1996a). Extensions to a Generalization Critic for Inductive Proof. In M. A. McRobbie, and J. K. Slaney, eds., *13th International Conference on Automated Deduction*. Springer-Verlag, pp. 47–61. Springer Lecture Notes in Artificial Intelligence No. 1104. Also available from Edinburgh as DAI Research Paper 786.

Ireland, A. and Bundy, A. (1996b). Productive use of failure in inductive proof. *Journal of Automated Reasoning*, **16**(1–2), 79–111. Also available from Edinburgh as DAI Research Paper No 716.

Ireland, A. and Bundy, A. (March 1999). Automatic Verification of Functions with Accumulating Parameters. *Journal of Functional Programming: Special Issue on Theorem Proving & Functional Programming*, **9**(2), 225–45. A longer version is available from Department of Computing and Electrical Engineering, Heriot-Watt University, Research Memo RM/97/11.

Ireland, A., Ellis, B.J., Cook, A., Chapman, R. and Barnes, J. (2004). An integrated approach to program reasoning. Under review by the Journal of *Formal Aspects of Computing: Special Issue on Integrated Formal Methods*, 2004. Available from the School of Mathematical and Computer Sciences, Heriot-Watt University, as Technical Report HW-MACS-TR-0027.

Ireland, A. and Stark, J. (1997). On the automatic discovery of loop invariants. In *Proceedings of the Fourth NASA Langley Formal Methods Workshop*. NASA Conference Publication 3356. Also available as Research Memo RM/97/1 from Department of Computing and Electrical Engineering, Heriot-Watt University.

Ireland, A. and Stark, J. (February 2001). Proof planning for strategy development. *Annals of Mathematics and Artificial Intelligence*, **29**(1–4), 65–97. An earlier version is available as Research Memo RM/00/3, Department of Computing and Electrical Engineering, Heriot-Watt University.

Ireland, A. (1992). The Use of Planning Critics in Mechanizing Inductive Proofs. In A. Voronkov, ed., *International Conference on Logic Programming and Automated Reasoning – LPAR 92, St. Petersburg*. Lecture Notes in Artificial Intelligence No. 624. Springer-Verlag, pp. 178–89. Also available from Edinburgh as DAI Research Paper 592.

Kaldewaij, A. (1990). *Programming: The Derivation of Algorithms*. London: Prentice Hall.

Katz, S. M. and Manna, Z. (1976). Logical analysis of programs. *Communications of the ACM*, **19**(4), 188–206.

King, S., Hammond, J., Chapman, R. and Pryor, A. (2000). Is proof more cost effective than testing? *IEEE Trans. on SE*, 26(8), 2000.

Kling, R. (1971). A paradigm for reasoning by analogy. *Artificial Intelligence*, **2**: 147–178.

Knuth, D. and Bendix, P. (1970). Simple word problems in universal algebras. In J. Leech, ed., *Computational Problems in Abstract Algebra*. Oxford: Pergamon Press, pp. 263–97.

Kolbe, T. and Walther, C. (1994). Reusing proofs. In A. G. Cohn, ed., *Proceedings of the Eleventh European Conference on Artificial Intelligence*. John Wiley and Son: Chichester.

Kolbe, T. and Walther, C. (1998). Proving theorems by reuse. *Artificial Intelligence*, **116**(1–2), 17–66.

Kowalski, R. (1979). *Logic for Problem Solving*. Artificial Intelligence Series, North Holland.

Kraan, I., Basin, D. and Bundy, A. (1996). Middle-out reasoning for synthesis and induction. *Journal of Automated Reasoning*, **16**(1–2), 113–45. Also available from Edinburgh as DAI Research Paper 729.

Manna, Z. and Waldinger, R. J. (1985). *The Logical Basis for Computer Programming*, Vol 1: *Deductive Reasoning*. Reading, Ma: Addison-Wesley.

McCune, W. (1991). OTTER 2.0 users guide. Technical report MCS-P220-0391, Argonne National Laboratory.

Melham, T. F. (1990). *Formalizing Abstraction Mechanisms for Hardware Verification in Higher Order Logic*. Unpublished Ph.D. thesis, Computer Laboratory, University of Cambridge.

Melis, E. (1995). A model of analogy-driven proof-plan construction. In *14th International Joint Conference on Artificial Intelligence*. Montreal: Morgan Kaufman, pp. 182–9.

Melis, E. and Schairer, A. (1998). Similarities and reuse of proofs in formal software verification. In B. Smyth, and P. Cunningham, eds., *Advances in Case-Based Reasoning, Proceedings 4th European Workshop on Case-Based Reasoning (EWCBR-98)*. Springer-Verlag, LNCS.

Morris, J. B. (1969). E-resolution: Extension of resolution to include the equality relation. In D. Walker, and L. M. Norton, eds., *Proceedings of IJCAI-69*. Kaufmann Inc., pp. 287–94.

Nadathur, G. and Miller, D. (1988). An overview of λProlog. In R. A. Kowalski and K. A. Bowen, eds., *Proceedings of the Fifth International Logic Programming Conference/Fifth Symposium on Logic Programming*. MIT Press.

Negrete, S. (June 1994). Guiding proof search in logical frameworks with rippling. In D. Galmiche and L. Wallen, eds., *Workshop on Proof Search in Type-Theoretic Languages*. Nancy, France: CADE, pp. 55–61. Also available from Edinburgh as DAI Research Paper No. 750.

Negrete, S. (May 1996). *Proof planning with logic presentations*. Unpublished Ph.D. thesis, Department of Artificial Intelligence, University of Edinburgh.

Newell, A., Shaw, J. C. and Simon, H. A. (1957). Empirical explorations with the Logic Theory Machine. In Proceeding Western Joint Competition Conference, pp. 218–39. Reproduced in *Computers and Thought* Feigenbaum and Feldman, eds., New York: McGraw Hill, pp. 109–33, (1963).

Nieuwenhuis, R. and Rubio, A. (1992). Theorem proving with ordering constrained clauses. In D. Kapur, (ed.), *Proceedings of the 11th International Conference on Automated Deduction (CADE-11)*. Saratoga Springs, NY: Springer-Verlag, pp. 477–91, LNAI 607.

Owen, S. (1990). *Analogy for Automated Reasoning*. Academic Press.

Pientka, B. and Kreitz, C. (1998). Automating inductive specification proofs in NuPRL. *Fundamenta Mathematicae*, **34**, 1–20.

Plaisted, D. (1980). Abstraction mappings in mechanical theorem proving. In *Proceedings of the Fifth International Conference on Automated Deduction*. Les Arcs, France: Springer, LNCS 87, pp. 264–80.

Prehofer, C. (1994). Higher-order narrowing. In *Proceedings of the 9th Annual IEEE Symposium on Logic in Computer Science*. IEEE Computer Society Press, pp. 264–80.

Quinlan, J. R. and Hunt, E. B. (1986). A formal deductive problem-solving system. *Journal of the Association for Computing Machinery*, **14**(4), 625–46.

Richardson, J. D. C, Smaill, A. and Green, I. (July 1998). System description: proof planning in higher-order logic with Lambda-Clam. In C. Kirchner, and H. Kirchner, eds., *15th International Conference on Automated Deduction*, vol. 1421 of *Lecture Notes in Artificial Intelligence*. Lindau, Germany: pp. 129–33.

Robinson, P. J. and Staples, J. (1993). Formalizing a hierarchical structure of practical mathematical reasoning. *Journal of Logic Computation*, **3**(1), 47–61.

Robinson, G. and Wos, L. (1969). Paramodulation and theorem proving in first order theories with equality. *Machine Intelligence*, **4**, 133–50.

Schairer, A. (1998). A technique for reusing proofs in software verification. Unpublished M.Sc. thesis, University of Stuttgart/DFKI.

Sengler, C. (1997). *Induction on Non-Freely Generated Data Types*. infix-Verlag, No. 160.

Smaill, A. and Green, I. (1996). Higher-order annotated terms for proof search. In J. von Wright, J. Grundy, and J. Harrison, eds., *Theorem Proving in Higher Order Logics: 9th International Conference*, TPHOLs 1996 vol. 1275 of *Lecture Notes in Computer Science*. Turku, Finland: Springer-Verlag, pp. 399–414. Also available as DAI Research Paper 799.

Staples, M. (1995). Window inference in isabelle. In *Proceedings of the Isabelle Users Workshop*, Cambridge: UK. University of Cambridge.

Stark, J. and Ireland, A. (1998). Invariant discovery via failed proof attempts. In P. Flener, ed., *Logic-Based Program Synthesis and Transformation*, no. 1559 in LNCS, Springer-Verlag, pp. 271–88. An earlier version is available from the Department of Computing and Electrical Engineering, Heriot-Watt University, Research Memo RM/98/2.

Velleman, J. D. (1994). *How to Prove it – A Structured Approach*. Cambridge: Cambridge University Press.

Walsh, T. (1994). A divergence critic. In A. Bundy, ed., *12th International Conference on Automated Deduction*, Lecture Notes in Artificial Intelligence, vol. 814. Nancy, France: Springer-Verlag, pp. 14–28.

Walsh, T., Nunes, A. and Bundy, A. (1992). The use of proof plans to sum series. In D. Kapur, ed., *11th International Conference on Automated Deduction*, Springer-Verlag, pp. 325–39. Lecture Notes in Computer Science No. 607. Also available from Edinburgh as DAI Research Paper 563.

Wikström, Å. (1987). *Functional Programming Using Standard ML*. Prentice-Hall.

Yoshida, T. Bundy, A. Green, I. Walsh, T. and Basin, D. (1994). Coloured rippling: An extension of a theorem proving heuristic. In A. G. Cohn, ed., *Proceedings of ECAI-94*. Chichester: John Wiley, pp. 85–9.

Index

Printed in the United States
by Baker & Taylor Publisher Services